AIR POLLUTION
AND
ATMOSPHERIC DIFFUSION

AIR POLLUTION
AND
ATMOSPHERIC DIFFUSION

M.E. Berlyand, Editor

Translated from Russian by A. Baruch

Translation edited by D. Slutzkin

A HALSTED PRESS BOOK

JOHN WILEY & SONS

New York · Toronto

ISRAEL PROGRAM FOR SCIENTIFIC TRANSLATIONS

Jerusalem · London

© 1973 Israel Program for Scientific Translations Ltd.

Sole distributors for the Western Hemisphere and Japan
HALSTED PRESS, a division of
JOHN WILEY & SONS, INC., NEW YORK

Distributors for the U.K., Europe, Africa and
the Middle East
JOHN WILEY & SONS, LTD., CHICHESTER

Distributed in the rest of the world by
KETER PUBLISHING HOUSE, JERUSALEM

Library of Congress Catalog Card Number 73 1982
ISBN 0 7065 1286 3 IPST, Jerusalem
ISBN 0 470 07034 X Halsted/Wiley, N.Y.
IPST cat. no. 22048

This book is a translation from Russian of
ATMOSFERNAYA DIFFUZIYA I
ZAGRYAZNENIE VOZDUKHA
Gidrometeoizdat, Leningrad 1971

Printed in Israel

Contents

GENERALIZATION OF THE THEORY OF THE DISPERSION OF INDUSTRIAL DISCHARGES IN THE ATMOSPHERE

M. E. Berlyand, R. I. Onikul

1. Statement of the problem

With the development of the theory of atmospheric diffusion of contaminants it became possible to formulate a new approach to the problem of preventing the pollution of the air basin, including fixing the permitted amount of harmful discharges in the atmosphere and the rational distribution of air purity observation stations. The theory can be used to recommend economically optimal sets of preventive measures, including the determination of the necessary degree of purification of smoke and ventilation gases, the selection of the height, diameter, and number of stacks, the location of industrial and residential areas, etc.

However, strict allowance for the laws of the dispersion of contaminants in the atmosphere in the calculation of the concentrations of harmful substances in the industrial area is not the only factor determining the success of the practical application of scientific recommendations. Such success is also very dependent on the accuracy of determination of the amount of discharge and of some other parameters in the computing formulas. The existing health criteria help to establish the degree of harmfulness of the calculated concentrations and to determine which air characteristics must be calculated.

Accordingly, studies have been started in the USSR for developing a method for calculating the dispersion of industrial discharges. Meteorogists, and specialists in atmospheric physics, technicians, and hygienists have played an active part in these studies. The studies started with the working out of a "Provisional method for calculating the dispersion of discharges in the atmosphere (sols and sulfur-containing gases) from the stacks of power plants" /19/, approved in 1963. This is the first standard for calculating the dispersion of discharges in the atmosphere; it was formulated for thermal power plants for a number of reasons.

Firstly, many of the powerful thermal power plants (TPP) discharge enormous quantities of sulfur dioxide into the atmosphere. A single 2,400 MW TPP operating on sulfur-containing mazut or hard coal discharges up to 1,000 tons daily. At the same time, the discharges from TPPs are never scrubbed for removing sulfur dioxide, since the cost of the purification installation is of the same order as the cost of the power plant itself. At the contemporary level of gas purification the only practical way out of this difficulty consists in building high stacks which will efficiently use the dispersive properties of the atmosphere. This requires the development of a scientific method for calculating the case of single high sources. With TPP as a simple type of source, the organization of experimental studies in the field of concentrations of the discharged contaminants was simplified, and a reliable test of the theoretical calculations could be developed.

Secondly, the methods for determining the amount of harmful substances formed during the burning of fuel have been developed much better than for most other industries.

Finally, the building of powerful TPP which will long remain the most important sources of electric power is extremely important. More than 80% of the world output of electric power is produced by TPP. The establishment of the computing method /19/ coincided with the initial stage of the design of a wide network of thermal power plants in the USSR in accordance with the twenty-year national electrification plan. The basic restrictions on the power of the designed power plants were necessary for ensuring the required purity of the air basin.

Earlier methods for calculating the dispersion of contaminants proved to be unsuitable in the above case, and unnecessarily hampered the development of power plants. The introduction of the procedure /19/ with the rational allowance for meteorological conditions stipulated by it, radically changed the situation: it raised by a factor of two the permissible limit of the TPP power, while ensuring air purity in the power plant area. The experimental material accumulated since the publication of paper /19/ has fully confirmed its conclusions.

The publication of paper /19/ was followed by the "Instructions for calculating the dispersion of harmful substances in the atmosphere (dust and sulfur dioxide) contained in the discharges of industrial enterprises" SN 369-67 /32/, approved by the USSR Gosstroi as compulsory specifications in the design and running of plants. These instructions extended the limits of applicability of the computing formulas, and gave methods for calculating concentrations from a group of sources and for determining the boundaries of the sanitary-protective zone.

The participation of specialists in gas purification, design of metallurgical plants, oil-processing plants, etc., in the development of the improved procedure, made it possible to work out standard methods for calculating the discharge parameters in the atmosphere for a number of important industries, and to extend the procedure to a relatively wide class of enterprises.

The basic formula used in /19/ and /32/ for calculating the maximum concentration of contaminants c_M from a single source is

$$c_M = \frac{A M_1 F m}{H^2 \sqrt[3]{V_1 \Delta T}}.$$
(1)

Here M_1 and V_1 are the amount of harmful substances and the volume of gases discharged by the stack per unit time, H is the height of the stack, ΔT is the difference in temperature between the discharged gases and the surrounding air, m is a dimensionless coefficient allowing for the exhaust conditions of the gas-air mixture, F a dimensionless coefficient allowing for the rate or precipitation of harmful substances in the atmosphere (for light discharges such as gaseous discharges, $F=1$). The coefficient A depends on the temperature stratification, determining the intensity of vertical and horizontal mixing in the atmosphere, and is taken into account under conditions of the least favorable combination of these factors. The value of c_M is related to the so-called dangerous wind velocity $u_M \approx v_M$ (m/sec), where

$$v_M = 0.65 \sqrt[3]{\frac{V_1 \Delta T}{H}}.$$
(2)

The formulas and some of the principles of the computing procedure described in /19/ and /32/ can be applied to widely different sources of discharge of harmful substances. However, the range of applicability of these first standards was limited, in particular because of the necessity of having a procedure for determining the parameters of discharge of the harmful substances. For the most important and polluting industries it was also considered advisable to first compare the calculated and empirical results to test the methods for calculating the concentrations and for determining the discharge parameters. The computing procedure was also limited to specific industries. Thus, paper /19/ was intended for a powerful and relatively hot discharge from high stacks, characteristic of power plants. The following conditions must hold if paper /19/ can be applied: $\Delta T \geqslant 30°$, $H \geqslant 50$, and $V_1 \geqslant 20$ m³/sec.

The limits of applicability were extended in /32/ by somewhat improving the computing formulas. These limits were estimated by a universal criterion, according to which the coefficient

$$f = 10^3 \frac{w_0^2 D}{H^2 \Delta T},$$

where D is the stack diameter, and w_0 the velocity of the discharged gases, and $f < 6 \, \text{m/sec}^2 \cdot \text{deg}$. However, the formulas for c_M and u_M are not applicable to cold discharges, when $\Delta T \approx 0$, since if ΔT decreases, the value of c_M increases without limit, the dangerous velocity u_M decreases to zero, and the parameter f increases as well. This most important restriction on the computing procedure given in /19/, namely the necessity of considerable overheating of the discharged gases, was retained in /32/, but it was partially removed by the introduction of the coefficient f.

However, many industries discharge unheated gases. These enter the atmosphere directly from the plants and are usually at a temperature differing little from that of the plant premises. Such are the conditions in the rayon fiber plants, where the carbon disulfide and hydrogen sulfide discharged are at approximately the same temperature as the premises.

When the theory for calculating the dispersion of contaminants is extended to the case of cold discharges, we must in the first place allow for the decrease in the dangerous wind velocities u_M. A more detailed examination shows that the extension of the method to small values of u_M makes it possible to widen appreciably the possibilities of its utilization, for sources with cool discharges, and also a wide class of other practically important sources.

It was suggested in /19/ and /32/ that the concentration c_M for heavy contaminants be calculated from formula (1) with F taken as 2 or 2.5, depending on whether the degree of dust removal from the stack gases is higher or lower than 90%. With this schematic approach the degree of dispersion of sols and dust and their specific weight, characteristic of TPP and some similar sources, can be approximately taken into account.

With the development of the computing method and the extension of the class of industries to which it is applicable, a more strict allowance must be made for the influence of the gravitational precipitation of the contaminant on its dispersion. We must take into account the existence of substantial differences in the spectral composition of the dust and in its density. Besides, as was shown in /12/, under given conditions the ground concentrations of light and heavy contaminants vary not only with the structural characteristics of the dust, but also with the source height. The influence of the source height is manifested mainly for sources more than 200—250 m high and becomes stronger with increase in the height of discharge. This is important, in view of the contemporary trend toward higher stacks and more powerful discharges.

The present paper discusses the solution of these questions: the extension of the computing method to the case of cold discharges and low dangerous velocities, and its improvement for heavy con- taminants. To develop a single approach to the calculation of the dispersion of industrial discharges in the atmosphere, we will try to extend the theoretical results in such a way that they will yield as particular cases the basic principles of the methods developed in /19/ and /32/.

2. Atmospheric diffusion of cold and heated gases

In our previous papers /7—10 etc./ we numerically solved the equation of turbulent diffusion of a contaminant from a high source:

$$u \frac{\partial q}{\partial x} + w \frac{\partial q}{\partial z} = \frac{\partial}{\partial z} k_z \frac{\partial q}{\partial z} + k_y \frac{\partial^2 q}{\partial y^2}. \tag{3}$$

Here q is the contaminant concentration, u the wind velocity, w the velocity of ordered displacement of the contaminant along the vertical, k_z and k_y the vertical and horizontal components of the eddy diffusivity. The x axis in the direction of the wind and the y axis perpendicular to it in the horizontal plane, and the z axis is in the vertical plane.

The problem was solved for different profiles of the wind velocity and the eddy diffusivity in the boundary layer of the atmosphere. We distinguished between normal, frequently met profiles, and ab- normal ones, observed much more rarely. The computing proce- dure for dispersion of contaminants used in designing measures for protecting the air basin from pollution must be based on normal conditions. Under such conditions the height profile of the wind velocity u is roughly logarithmic, while the vertical component k_z of the eddy diffusivity increases linearly with height in the ground layer (of height h), and remains roughly constant for $z > h$. Thus, we can write

$$u = u_1 \frac{\ln \frac{z}{z_0}}{\ln \frac{z_1}{z_0}}, \tag{4}$$

$$k_z = \begin{cases} \nu + k_1 \dfrac{z}{z_1} & \text{for } z \leqslant h \\[2ex] \nu + k_1 \dfrac{h}{z_1} & \text{for } z > h. \end{cases} \qquad (5)$$

Here ν is the coefficient of molecular diffusion for air, u_1 and k_1 are the values of u and k_z respectively, for $z = z_1$ ($z_1 = 1m$), z_0 is the roughness of the underlying surface.

The allowance for some deviations from (4) and (5) in the numerical solution of (3) presents no difficulties, but as shown in /7/ etc., this allowance is of no appreciable importance in the determination of the ground concentration, and by neglecting the standardization of the computing procedure is considerably simplified.

The concentrations obtained by integrating (3) were averaged over the time period T (/6, 9/). T is usually equal to 20 min., corresponding to the time period to which the single values of the ultimate admissible concentrations are related. In the averaging it was taken into account that the wind varies in the horizontal plane, with a probability $\omega(\varphi)$ of deviation in its direction by an angle φ from its mean direction:

$$\omega(\varphi) = \frac{1}{\varphi_0 \, 2\sqrt{\pi}} \, e^{-\frac{\varphi^2}{2\varphi_0^2}}, \qquad (6)$$

where φ_0 is the variance of the wind direction in time T.

This averaging is equivalent to the introduction of an effective value of the horizontal component k_y of the eddy diffusivity, such that $k_y \approx \varphi_0^2 u x$ for sufficiently large values of x. According to /12/, the average value of the ground concentration of a light gaseous contaminant ($w \approx 0$) at a height H of the source is expressed in a general form by the formula

$$q = q_m \, \Phi_{1H}\left(\frac{x}{x_m}\right) e^{-\frac{y^2}{2\varphi_0^2 x^2}}, \qquad (7)$$

where q_m is the maximum value of q, reached when $x = x_m$ and $y = 0$; Φ_{1H} is some function determined in the numerical solution of the problem for the given value of H.

The values of q_m and x_m are approximately determined from the relations:

$$q_m = \frac{C_1 M k_1}{u_1^2 \varphi_0} \frac{1}{H^{\beta_1}},$$

$$x_m = C_2 \frac{u_1}{k_1} H^{1 + \beta_2}. \qquad (8)$$

Here M is the output of the discharge, and C_i and β_i $(i = 1, 2)$ are constants which are little dependent on h and z_0. Usually $\beta_1 \approx 2, 3$ and $\beta_2 \approx 0.2$.

Industrial discharges into the atmosphere have a specific exit velocity (w_0) from the stack, and if the gases are at a temperature (ΔT) above that of the surrounding air, they possess buoyancy as well. As a result, a field of vertical velocities is created near the source, that contributes to the initial rise of the discharged particles. By solving the system of equations describing the field of velocities appearing at the jet in the presence of wind, and the turbulent diffusion of the heat from the stack, we can approximately determine the effective rise ΔH. According to /6, 9/,

$$\Delta H = \frac{1.5 w_0 R_0}{u}\left(2.5 + \frac{3.3 g R_0 \Delta T}{T_e u^2}\right), \tag{9}$$

where u is the wind velocity at the wind vane level, R_0 is the radius of the stack outlet, g the gravitational acceleration, and T_e the absolute temperature of the surrounding air.

The dependence of the maximum ground concentration of the contaminant q_m on the wind velocity u is complex. As seen from (8), for a fixed source height H the value of q_m is inversely proportional to u. At the same time, when u increases, ΔH and the effective source height $H + \Delta H$ decrease, and as a result q_m and q increase. The dangerous wind velocity at which the ground concentration of the contaminant becomes maximum is determined from the condition

$$\frac{\partial q_m}{\partial u} = 0. \tag{10}$$

It is necessary to take into account not only the above dependences of q_m and ΔH on u described by formulas (8) and (9), but also the variation in the combination of parameters $K = \frac{k_1}{u_1 \varphi_0}$ given in (8) with it.

According to contemporary concepts, the ratio $\frac{k_1}{u_1}$ depends mainly on the atmospheric stability index $B = \frac{\delta T}{u_1^2}$, where $\delta T = T_{z_2} - T_{z_3}$ is the temperature difference at two levels in the ground layer of air (usually $z_2 = 0.5$ m and $z_3 = 2$ m). According to /20/ and /25/, the value of φ_0 also depends on B. The values of $\frac{k_1}{u_1}$ and φ_0 increase with increase in instability, characterized by increase in B. In the case of stable stratification of the ground air layer, increase in the modulus B is accompanied by a decrease in the value of $\frac{k_1}{u_1}$, and φ_0 displays a weak tendency to increase from some minimum value reached under conditions near the equilibrium stratification of air (B = 0). As a result, K decreases at inversion with the increase in the stability of the atmosphere. Therefore, in the case of a super-adiabatic gradient there will be a higher maximum ground concentration of a contaminant from a high source than at inversion, due to the proportionality between C and K. It follows that to determine the maximum possible value of the ground concentration and to estimate the dangerous wind velocity corresponding to it, it suffices to consider the dependence of K on u for unstable convective conditions. In /6/ and /12/, the authors used the fact that under these conditions at $u_1 > 2$ m/sec, the value of K is little dependent on u_1.

It follows from the above that we must examine here in greater detail the dependence of $\frac{k_1}{u_1 \varphi_0}$ on u_1, in particular in the region of weak winds. The characteristic features of atmospheric diffusion in the limiting case of a weak wind, near to calm, were investigated in /14/ taking into consideration dimensions and the known results on turbulence at free convection, when there is no directed air transfer. Under such conditions the vertical and horizontal components k_z and k_y of the eddy diffusivity are practically independent of the wind velocity, and according to /29/, k_z and $k_y \sim \sqrt{\delta T}$, where δT is the temperature difference at two levels in the ground layer.

It was accordingly assumed in /14/ (with the aid of the expression given above for the effective horizontal component of eddy diffusivity $k_y \approx \varphi_0^2 ux$), that $\frac{k_1}{u_1} = \frac{\alpha}{u_1}$, and $\varphi_0 = \frac{\beta}{\sqrt{u_1}}$, where α and β are constants, and $\alpha \sim \sqrt{\delta T}$, $\beta \sim \sqrt[4]{\delta T}$. Thus, in the limiting case of weak winds, $K = \frac{a_1}{\sqrt{u_1}}$, where a_1 is a constant magnitude little dependent on $\delta T (a_1 \sim \sqrt[4]{\delta T})$.

In general we examined the dependence of K on u_1 on the basis of the results of the separate investigation of the dependence of $\frac{k_1}{u_1}$ and φ_0 on u_1.

A number of formulas are available for determining the eddy diffusivity k_1, and therefore $\frac{k_1}{u_1}$ from the results of gradient obser-vations. In the case of unstable stratification, similar results are usually obtained. The differences between the results do not exceed 30%, which is within the limits of the accuracy of the determination of k_1. This follows, for example from a comparison between the widely used procedures of Budyko /15/ and of Kazanskii and Monin /28/.

The influence of the stability parameter B on the eddy diffusivity is clearly seen from Budyko's formula, which can be written in the following form in the case of a logarithmic height profile of the wind velocity and the temperature:

$$k_1 = k_{1p}\sqrt{1 + \gamma B},$$

where $k_{1p} = \dfrac{\varkappa^2 u_1}{\ln \dfrac{z_1}{z_0}}$, $\varkappa$ is the von Karman constant, $\gamma = 15\,\mathrm{m}^2/\mathrm{sec}^2 \cdot \deg$ at $z_0 = 0.01\,\mathrm{m}$. It follows from this formula that in the limiting case of free convection (at $u \approx 0$) and $z_0 = 0.01\,\mathrm{m}$, $k_1 \approx 0.15\sqrt{\delta T}$. This result agrees satisfactorily with those for free convection obtained by considering both the theory and the experimental data.

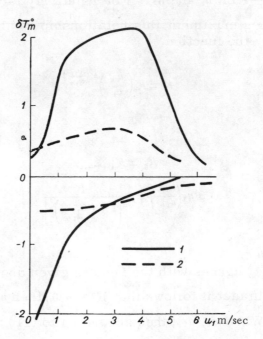

FIGURE 1. δT_m as a function of u_1 in August (1) and February (2).

For a fixed value of u_1, the maximum values of k_1 are obtained for the maximum value of $\delta T = \delta T_m$. It should be remembered that u_1 and δT_m are correlated. As an example, Figure 1 shows a graph plotted by Gracheva /26/, representing δT_m as a function of u_1 from the data of observations at the Rudnya meteorological station in 1967. The upper part of the graph corresponds to unstable stratification, and the lower part to inversion conditions. It is seen that in the case of unstable stratification δT_m is maximum at a wind speed $u_1 = 2-3$ m/sec, and in the case of stable stratification under calm conditions.

The dependence of φ_0 on u_1 and δT was examined by processing the results of a continuous recording of the wind direction, carried out systematically by the GGO expeditions studying air pollution. The results of an analysis of a large number of the data are described in /20/ and /25/. It follows from these that in the case of unstable stratification and fixed $\delta T, \varphi_0$ decreases with increase in u_1, while at a given u_1 it increases somewhat with increase in δT.

From these results for $\frac{k_1}{u_1}$ and φ_0 we can determine the dependence of $K = \frac{k_1}{u_1 \varphi_0}$ on u_1 and δT. For relatively large values of δT, when K becomes maximum, this relationship can be satisfactorily interpolated by the function

$$K = K_0 \left(\frac{u_{10}}{u_1} \right)^{n_1 \left(\frac{u_1}{u_{10}} \right)}, \tag{11}$$

where

$$K_0 = K|_{u_1 = u_{10}},$$

$$n_1 \left(\frac{u_1}{u_{10}} \right) = \frac{1}{2 \left[1 + \left(\frac{2u_1}{u_{10}} \right)^2 \right]}, \tag{12}$$

and $u_{10} = 2$ m/sec.

Formula (11) agrees with the results given above for the case of weak winds. In fact it follows that if $u_1 \to 0$, then $n_1 \to \frac{1}{2}$ and $K = K_0 \sqrt{\frac{u_{10}}{u_1}}$. When $u_1 > u_{10}$, then according to (11), K differs little from K_0.

We shall now use condition (10) allowing for (4), (8), (9), and (11), to find the dangerous velocity u_M under unsatisfactory meteorological conditions, and the corresponding value of the maximum concentration q_M. It is convenient to introduce the parameters v_M and $f = 10^3 \frac{w_0^2 D}{H^2 \Delta T}$ mentioned in the preceding sections. These parameters

will play an important role in the part below.   It is therefore advisable to find the possible range of their variation for real sources.

For sufficiently powerful sources of heated discharges TPP (heat
and electric power plants, sintering works, cement plants, etc.) at
$H = 50-250$ m, $\Delta T = 50-200°$, $w_0 = 10-30$ m/sec, $D = 3-7$ m, $V_1 = 50-$
$1,200$ m^3/sec.   If we consider the relationships between these parameters, the values of v_M are relatively large and vary between
$2-7$ m/sec, while f is as a rule $0.5-2$ m/sec$^2 \cdot$ deg, at times reaching
$5-6$ m/sec$^2 \cdot$ deg.

For slightly heated discharges, and also at a low power of the
sources, as in the case of most ventilation discharges, at $H = 20-50$ m,
$\Delta T = 5-10°$, $w_0 = 2-5$ m/sec, $D = 0.5-2$ m, and $V_1 = 1-10$ m^3/sec,
v_M usually varies within $0.5-1$ m/sec, and f is greater than
10 m/sec$^2 \cdot$ deg.

The parameters v_M and f are correlated; as a rule, a small v_M
corresponds to a large f, and a large v_M to a small f.   Using v_M and
f, we can write formula (9) for ΔH in the form

$$\Delta H = H\left(\frac{0.2 v_M \sqrt[3]{f}}{u_1} + 0.05 \frac{v_M^3}{u_1^3}\right). \tag{13}$$

The following equation is then derived from (10) for determining the
dangerous velocity $u_1 = u_{1M}$:

$$\frac{d}{du_1}\left[u_1\left(\frac{u_1}{u_{10}}\right)^{n_1\left(\frac{u_1}{u_{10}}\right)}\left(1 + 0.2\frac{v_M\sqrt[3]{f}}{u_1} + 0.05\frac{v_M^3}{u_1^3}\right)^{2,3}\right] = 0. \tag{14}$$

It can be transformed to

$$\left(\frac{v_M}{u_{1M}}\right)^3 + 4\left(\frac{t - 1.3}{t - 5.9}\right)\sqrt[3]{f}\frac{v_M}{u_{1M}} + 20\frac{(1+t)}{(t - 5.9)} = 0, \tag{15}$$

where

$$t = n_1\left(\frac{u_{1M}}{u_{10}}\right) + u_{1M}\ln\left(\frac{u_{1M}}{u_{10}}\right)\frac{d}{du_1}\left[n_1\left(\frac{u_{1M}}{u_{10}}\right)\right]. \tag{16}$$

By solving the cubic equation (15) with respect to v_M, we obtain

$$v_M = u_{1M}\sqrt[3]{\frac{10(1+t)}{5.9 - t}}\left(\sqrt[3]{1 + \sqrt{1+\varphi}} + \sqrt[3]{1 - \sqrt{1+\varphi}}\right), \tag{17}$$

where

$$\varphi = 0.024 f\frac{(t - 1.3)^3}{(1+t)^2(t - 5.9)}. \tag{18}$$

The dependence of u_{1M} on v_M and f can be found graphically. However, this necessitates laborious calculations.   We used a more convenient procedure, consisting in solving (14) numerically on a computer.

In accordance with (4), the value of the dangerous velocity u_{1M} at the wind vane level z is directly related to the value of u_M.

The calculations gave the dependence of the ratio $\frac{u_M}{v_M}$ on f for different values of v_M.   Two cases are considered: $v_M < u_{10}$ and $v_M > u_{10}$.

In the first case it follows from the dependence of $\frac{u_M}{v_M}$ on f for different values of v_M that $u_M \approx v_M$ over a wide range of variation in f and v_M .

In the second case $\frac{u_M}{v_M}$ is little dependent on v_M; the values of u_M and v_M are similar at small f values only, but the ratio $\frac{u_M}{v_M}$ appreciably increases with increase in f.

We can write approximately

$$u_M = v_M \quad \text{for} \quad v_M < 2\,\text{m/sec.,}$$
$$u_M = v_M \left(1 + 0.12\sqrt{f}\right) \quad \text{for} \quad v_M > 2\text{ m/sec.} \tag{19}$$

For the values of u_M obtained, the expression for the maximum concentration q_M can be written in the form

$$q_M = C_3 K_0 \, \frac{Mm}{H^2 \sqrt[3]{V_1 \, \Delta t}} \, E_1(v_M, \; f), \tag{20}$$

where

$$E_1(v_M, f) = \frac{\left(\dfrac{u_{10}}{u_{1M}}\right)^{n_1 \left(\frac{u_{1M}}{u_{10}}\right)}}{G_2(f)\left[1 + 0.2\,\dfrac{v_M \sqrt[3]{f}}{u_{1M}} + 0.05\,\dfrac{v_M^3}{u_{1M}^3}\right]^{2.3}}, \tag{21}$$

$$C_3 = \frac{1{,}54 C_1 \ln \dfrac{z_\phi}{z_0} \; G_2(f_0)}{\ln \dfrac{z_1}{z_0}}, \tag{22}$$

$$m = \frac{G_2(f)}{G_2(f_0)}, \tag{23}$$

$f_0 = 0.42 \text{ m/sec}^2 \cdot \text{deg}$ corresponds to combinations of the discharge parameters H_1, w_0, D, and ΔT frequently found in large industrial and power plants (for example $H = 120 \text{ m}$, $D = 6 \text{ m}$, $w_0 = 10 \text{ m/sec}$, $\Delta T = 100°$).

The values of G_2 for different values of f are given in Table 1.

TABLE 1.

f (m/sec²·deg) ...	0	0.42	1	5	10	50	100	1,000
$G_2(f)$	0.70	0.47	0.41	0.35	0.27	0.18	0.14	0.07

Calculations show that at $v_M < 2 \text{ m/sec}$, $E_1(v_M, f)$ depends mainly on v_M. At $v_M \geqslant 2-3 \text{ m/sec}$, $E_1 \approx 1$.

The distance x_M at which the value q_M is attained is found according to (8) and (13) from the formula

$$x_M = C_4 \left(\frac{u_{1M}}{u_{10}} \right)^{n_1 \left(\frac{u_{1M}}{u_{10}} \right)} H^{1.2} E_2(v_M, f), \qquad (24)$$

where

$$C_4 = 10 C_2,$$

$$E_2(v_M, f) = \left(1 + \frac{0.2 v_M^3 \sqrt{f}}{u_{1M}} + 0.05 \frac{v_M^3}{u_{1M}^3} \right)^{1.2} \sqrt{u_{1M}}. \qquad (25)$$

In the derivation of formula (24), the ratio $\frac{u_1}{k_1}$ at the dangerous wind velocity and unfavorable meteorological conditions was approximated by an analytical function analogous to (11):

$$\frac{u_1}{k_1} = \left(\frac{u_1}{k_1} \right)_{u_1 = u_{10}} \left(\frac{u_1}{u_{10}} \right)^{n_1 \left(\frac{u_1}{u_{10}} \right) + \frac{1}{2}},$$

If $u \neq u_M$, we can find as above expressions for the maximum concentration at this wind velocity and an unfavorable stratification q_{Mu} and for the corresponding distance x_{Mu}:

$$q_{Mu} = q_M E_3(v_M, f, u_1);$$
$$x_{Mu} = x E_4(v_M, f, u_1), \qquad (26)$$

where

$$E_3(v_M, f, u_1) = \frac{\left(\frac{u_{10}}{u_1}\right)^{n_1 \left(\frac{u_1}{u_{10}}\right)} \cdot u_{1M}}{\left(\frac{u_{10}}{u_{1M}}\right)^{n_1 \left(\frac{u_{1M}}{u_{10}}\right)} u_1} \left(\frac{1 + 0.2 \frac{v_M}{u_{1M}} \sqrt[3]{f} + 0.05 \frac{v_M^3}{u_{1M}^3}}{1 + 0.2 \frac{v_M}{u_1} \sqrt[3]{f} + 0.05 \frac{v_M^3}{u_1^3}}\right)^{2.3}, \tag{27}$$

$$E_4(v_M, f) = \sqrt{\frac{u_1}{u_{1M}}} \frac{\left(\frac{u_1}{u_{10}}\right)^{n_1 \left(\frac{u_1}{u_{10}}\right)}}{\left(\frac{u_{1M}}{u_{10}}\right)^{n_1 \left(\frac{u_{1M}}{u_{10}}\right)}} \left(\frac{1 + 0.2 \frac{v_M}{u_1} \sqrt[3]{f} + 0.05 \frac{v_M^3}{u_1^3}}{1 + 0.2 \frac{v_M}{u_{1M}} \sqrt[3]{f} + 0.05 \frac{v_M^3}{u_{1M}^3}}\right)^{1.2}. \tag{28}$$

Obviously, when $u_1 = u_{1M}$, the quantities E_3 and E_4 are equal to one.

The formulas (20), (24), (26) hold over a wide range of variation in u, v_M, and f. They can be used in calculations for relatively cold discharges, when $v_M < 0.5$ m/sec, and f reaches 100 m/sec$^2 \cdot$ deg or more. At the same time, it is expedient to consider the limiting case of cold discharges at $\Delta T = 0$, when $v_M = 0$, and $f \to \infty$. It is therefore convenient to consider again the formula for ΔH (9), substituting $\Delta T = 0$, Then

$$\Delta H = \frac{bH}{u_1}, \tag{29}$$

where

$$b = 2.5 \frac{w_0 R_0}{H}.$$

From (10), in the same way as above, we obtain an equation for finding the dangerous wind velocity $u_1 = u_{1M}$

$$\frac{d}{du_1}\left[u_1 \left(\frac{u_1}{u_{10}}\right)^{n_1 \frac{u_1}{u_{10}}} \left(1 + \frac{b}{u_1}\right)^{2.3}\right] = 0. \tag{30}$$

This equation can be transformed to the form

$$u_{1M} \frac{1+t}{1.3-t} = b, \tag{31}$$

where t is determined as before from formula (16). We then obtain

$$q_M = \frac{C_1 K_0 \left(\frac{u_{10}}{u_{1M}}\right)^{n_1 \left(\frac{u_{1M}}{u_{10}}\right)} M}{u_{1M} H^{2.3} \left(1 + \frac{b}{u_{1M}}\right)^{2.3}}, \tag{32}$$

$$x_M = C_4 \sqrt{u_{1M}} \left(\frac{u_{1M}}{u_{10}}\right)^{n_1 \left(\frac{u_{1M}}{u_{10}}\right)} H^{1.2} \left(1 + \frac{b}{u_{1M}}\right)^{1.2}.$$

Formulas (31) and (32) can be reduced to a simple form at sufficiently low velocities u_{1M}: $n_1 \simeq \frac{1}{2}$, and also when $u_{1M} > u_{10}$: $n_1 \approx 0$. It follows from (16) that in these cases $t = \frac{1}{2}$ and $t = 0$, respectively. For low dangerous wind velocities we then obtain for u_M at the wind vane level

$$u_M \simeq 0.8b, \qquad (33)$$

and for relatively high dangerous speeds

$$u_M \approx 2.0b. \qquad (34)$$

The corresponding expressions for q_M and x_M at low dangerous air velocities have the following form:

$$q_M = \frac{C_5 \sqrt{u_{10}} K_0 M}{H^{0.8} (w_0 R_0)^{1.5}},$$

$$x_M = C_6 \frac{w_0 R_0}{\sqrt{u_{10}}} H^{0.2}, \qquad (35)$$

where $C_5 = 0.086 \, C_1$; $C_6 = 48 \, C_2$.

For relatively high dangerous wind velocities ($u_M > 3 \, \text{m/sec}$), which are usually found when $v_M > 2 \, \text{m/sec}$

$$q_M = \frac{C_7 K_0 M}{w_0 R_0 H^{1.3}},$$

$$x_M = C_8 H^{0.7} \sqrt{w_0 R_0}, \qquad (36)$$

where $C_7 = 0.126 \, C_1$; $C_8 = 36 \, C_2$

It is interesting to compare the results of the calculations of q_M and u_M from the general formulas for large values of f and from the formulas for cold discharges. In the formula for ΔH we used as output parameters b and f instead of V_M and f:

$$\Delta H = H \left(\frac{b}{u_1} + \frac{6.25 b^3}{f u_1^3} \right).$$

Figure 2 represents u_{1M} as a function of b for $f = 100 \, \text{m/sec}^2 \cdot \text{deg}$ and $f = 1{,}000 \, \text{m/sec}^2 \cdot \text{deg}$, and also for cold discharges, calculated from (31). The broken lines on this figure represent the limiting values of u_{1M} as a function of b corresponding to formulas (33) and (34).

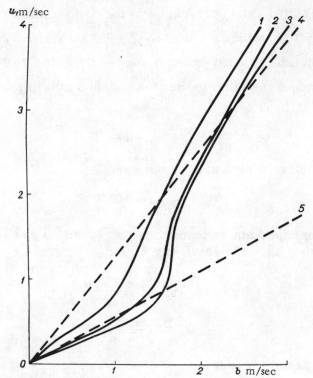

FIGURE 2. Dangerous wind velocity u_{1M} as a function of parameter b:

1) $f = 100\,\text{m/sec}^2 \cdot \text{deg}$; 2) $f = 1,000\,\text{m/sec}^2 \cdot \text{deg}$; 3) cold discharges, calculated from formula (31); 4) from formula (34); 5) from formula (33).

The differences in u_{1M} between heated and cold discharges are small at $f = 100\,\text{m/sec}^2 \cdot \text{deg}$, while at $f = 1,000\,\text{m/sec}^2 \cdot \text{deg}$ the dangerous velocities practically coincide in the two cases.

The calculations showed that the differences in the values of the maximum concentrations q_M at $f = 100\,\text{m/sec}^2 \cdot \text{deg}$ and for cold discharges do not exceed 10—20% over the entire possible range of variation in parameter b.

We should note that large values of f usually correspond to the conditions of low values of ΔT, when the exact determination of the latter does not seem possible in practice. We shall examine, as an example, the two extreme cases of high and low power of the source. We take for the first case $H = 150\,\text{m}$, $D = 6\,\text{m}$, $w_0 = 20\,\text{m/sec}$, and for the second case $H = 20\,\text{m}$, $D = 1\,\text{m}$, and $w_0 = 5\,\text{m/sec}$. Then, for powerful sources $f = 100\,\text{m/sec}^2 \cdot \text{deg}$ corresponds to $\Delta T = 1°$, and $f = 1,000\,\text{m/sec}^2 \cdot \text{deg}$ to $\Delta T = 0.1°$. The corresponding values for low-power sources are 0.6° and 0.06°. The differences in the gas temperature for both values of f are insignificant, and are practically impossible to determine. It follows that the general formulas (19), (20), (24) and (26) can be extended to cold discharges, when ΔT is equal to zero. In this case it is sufficient in practice to transform the computing formulas, and substitute in them $f = 100 \text{ m/sec}^2 \cdot \text{deg}$ and the value $\Delta T = \dfrac{10\,w_0^2 D}{H^2}$ corresponding to it.

We now proceed to the transformation of the computing formulas to a more convenient form.

The basic criterion of the degree of air pollution in an industrial district is the maximum value of the possible concentration c_M.* According to health requirements, its value should not exceed the single value of the maximum allowable concentration (MAC).

In the case of a light gaseous contaminant the expression for c_M is found from (20) by converting it to a form similar to (1):

$$c_M = \frac{AMmn}{H^2 \sqrt[3]{V_1 \Delta T}}. \tag{37}$$

Here, as in /5/ and /11/,

$$A = 1.54 C_1 \frac{\lg \dfrac{z_\phi}{z_0}}{\lg \dfrac{z_1}{z_0}} G_2(f_0) K_0 \tag{38}$$

or $A = a K_0$, where a is a constant approximately equal to 0.3.

For the factor m, depending on f, we use the same normalization as in /12/, according to which $m = 1$ at $f = f_0$.

Calculations from formula (21) showed that E_1 is very dependent on v_M but much less on f. Therefore, in the formula for c_M a new coefficient n is introduced, representing the value of E_1 averaged with reference to f.

The value of K_0 contained in A corresponds to the maximum values of δT at wind velocity $u_{10} = 2$ m/sec, and an unstable stratification. At $u_1 = u_{10}$ this temperature difference δT reaches values near the maximum δT_m, which corresponds to maximum values of $\dfrac{k_1}{u_1}$.

The value of δT_m, mentioned above, varies throughout the year at a given place (Figure 1). Besides, the value of δT_m depends on the physicogeographical conditions and the regional zones determined

---

* The magnitude c is used instead of q for the concentration because c is more widely utilized in practical calculations of the atmospheric pollution.

by them.   Values of the yearly variation in δT_m for a large number
of stations located in different regional zones of the USSR (taiga,
forest-steppe, steppe, semidesert, etc.), are given in /26/.   The
values of δT_m usually increase with transition from taiga to steppe
and from steppe to semidesert.   If we take the above into consideration, the values of A for different geographical regions are
determined as in /18/ and /32/.

If c_M is expressed in mg/m³, M in g/sec, H in m, V_1 in m³/sec,
ΔT in degrees, then $A = 120$ for open flat country in the center of the
European territory of the USSR, and in regions with similar climatic
conditions.   For the warmest regions of the USSR and forest regions
characterized by intense turbulent exchange, $A = 200$.   For regions
with intermediate conditions of turbulent mixing, A is taken as 160.

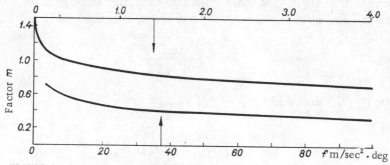

FIGURE 3.   Factor m as a function of parameter f.

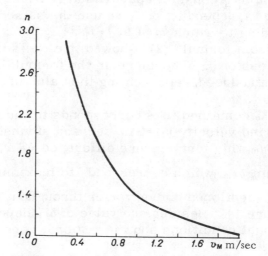

FIGURE 4.   Factor n as a function of parameter v_M.

The coefficient m is a function of a single argument f, and is determined from the curve plotted in Figure 3. When $f < 6\,\mathrm{m/sec^2 \cdot deg}$, the values of m coincide roughly with those given in /32/.

The main difference between the basic formula (37) and formula (1), used in /19/ and /32/, consists in the presence of the dimensionless factor n. The dependence of this factor on the parameter v_M is plotted in Figure 4. When $v_M \geqslant 2\,\mathrm{m/sec}$, n is taken as one, and then formula (37) reduces to the widely applied formula (1).

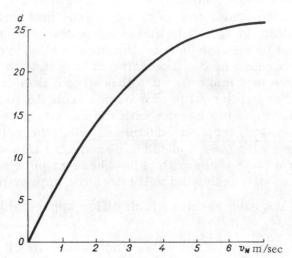

FIGURE 5. Factor d as a function of parameter v_M.

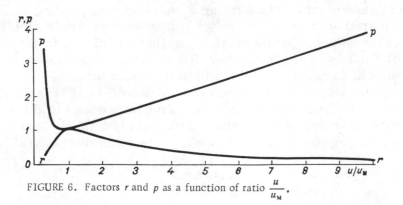

FIGURE 6. Factors r and p as a function of ratio $\dfrac{u}{u_M}$.

The value of the dangerous velocity u_M is determined from formula (19), and in these cases $v_M < 2\,\mathrm{m/sec}$ and $v_M > 2\,\mathrm{m/sec}$ are treated separately. However, at a very small v_M, less than

0.5 m/sec, the calculation of u_M from (21) is meaningless, since such low wind velocities are not measured in practice and are rarely observed at unstable stratification. It is therefore advisable to write $u_M \approx 0.5$ m/sec for $v_M < 0.5$ m/sec.

The value of c_M, determined from formula (37), corresponds to the distance

$$x_M = dH. \qquad (39)$$

The dimensionless coefficient d is determined approximately from (24), under the same conditions as those for the determination of c_M; this depends mainly on v_M. It is therefore possible to average its values over f, and to replace $H^{0.2}$ by the mean value, which varies little over a wide range of H. The value of d is determined from v_M by using the curve in Figure 5. It follows from this curve that at a sufficiently large v_M, the value of d varies from 15 to 25, i.e., on the average $x_M = 20H$. This agrees with the results given in /19/ and /32/. At $v_M > 2$ m/sec, and $f < 6$ m/sec$^2 \cdot$ deg, the value $u_M \approx v_M$ which also agrees with /19/ and /32/. When $v_M > 2$ m/sec, $n \approx 1$, and the expression for c_M agrees with (1). These results seem very reasonable, since with such wind velocities the approximations used for K and $\frac{u_1}{k_1}$ do not give results which differ appreciably from previously used values.

When $v_M < 2$ m/sec, the values of c_M and x_M are much more accurate.

Thus, the formulas for c_M, u_M, and x_M obtained in this paper are convenient, as we obtain the earlier results, corresponding to heated contaminants and relatively high dangerous wind velocities as particular cases. At the same time, they enable us to extend the method of calculation of the dispersion of contaminants over a much wider range of variation in the emission parameters.

In the same way as we obtained the values of c_M (37) from q_M (20), and x_M (39) from (24) when $u = u_M$, we can proceed to the maximum value of the concentration under unfavorable conditions $c_{M\,u}$ and the corresponding value $x_{M\,u}$ when $u \neq u_M$ from formulas (26), (27), (28):

$$c_{M\,u} = c_M r;$$
$$x_{M\,u} = x_M p. \qquad (40)$$

The values of r and p depend in general on v_M, f, and u. It is convenient to introduce $\frac{u}{u_M}$ as a new argument. In the case of weak sources with low values of v_M, relatively large values of $\frac{u}{u_M}$ are of practical interest; the opposite occurs with powerful sources with large values of v_M, then $\frac{u}{u_M}$ is small. If we also take into account that, depending on the source output, we have a definite correlation between v_M and f, containing essentially the same parameters, we can approximately reduce r and p to functions of one argument u/u_M. These functions are represented graphically in Figure 6.

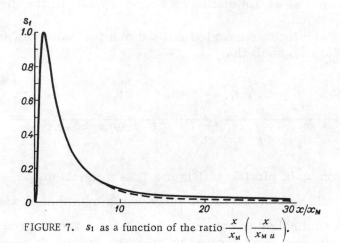

FIGURE 7. s_1 as a function of the ratio $\dfrac{x}{x_M}\left(\dfrac{x}{x_{M\,u}}\right)$.

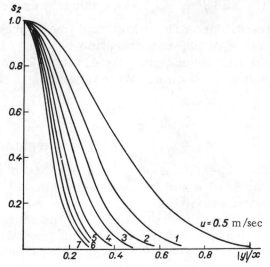

FIGURE 8. s_2 as a function of the wind speed and the ratio $\dfrac{|y|}{x}$.

It is interesting that r and p depend differently on $\frac{u}{u_M}$, as seen in Figure 6. When $\frac{u}{u_M} < 1$, r increases with increase in $\frac{u}{u_M}$, and there-fore $\frac{u}{c_{Mu}}$ increases and approaches c_M; p and x_{Mu} also approach x_M. When $\frac{u}{u_M} > 1$, the increase in $\frac{u}{u_M}$ leads to a decrease in r and an increase in p. This means that as we move away from the values c_M and x_M, the value of c_{Mu} decreases, and x_{Mu} increases.

Under the same conditions as those under which c_M and c_{Mu} were determined, it is possible to find the maximum value of the con-centrations c and c_u at distances $x \neq x_M$ and $x \neq x_{Mu}$ in the direction of the average wind.

From the calculations carried out we can introduce an inter-polation function s_1, such that

$$c - c_M s_1 \left(\frac{x}{x_M}\right) \quad \text{for} \quad u = u_M;$$

$$c = c_{Mu} s_1 \left(\frac{x}{x_{Mu}}\right) \quad \text{for} \quad u \neq u_M.$$

(41)

The function s_1 is plotted in Figure 7 as a function of $\frac{x}{x_M}$ and $\frac{x}{x_{Mu}}$ When $\frac{x}{x_M} > 8$ the concentration of gaseous substances is calcu-lated from the unbroken curve, and of dust from the broken curve. The relationship has been plotted up to much higher values of $\frac{x}{x_M} \left(\frac{x}{x_{Mu}}\right)$ than in /19/ and /32/, as it was sometimes necessary to superpose the concentrations from high and low sources.

The maximum value of the concentration c_y along the y axis at a distance x is determined according to (41) and (7), taking into account the dependence of φ_0 on u. We can write

$$c_y = c s_2.$$

(42)

Here s_2 is a function of $\frac{|y|}{x}$ and u. It is represented in Figure 8, which shows that the dependence of s_2 on u is more marked for relatively small values of u and is much less marked for the range of relatively high wind velocities.

It was mentioned above that calculation by the formulas given when $f = 100 \, \text{m/sec}^2 \cdot \text{deg}$ is suitable in practice for cold discharges, at the same temperature as the surrounding air ($\Delta T = 0$).

To analyse the influence of the operating factors, it is convenient to transform the computing formulas for $f = 100 \text{ m/sec}^2$, and eliminate ΔT from them. In formula (37) we must substitute for c_M and in formula (2) for v_M the value $\Delta T = \Delta T_0$ obtained from the formula for f, that is $\Delta T = \dfrac{10\, w_0^2 D}{H^2}$. The use of smaller values of ΔT to make the calculations more accurate is useless in practice. It then follows from (37) that

$$c_M = \frac{AMnl}{H^{4/3}},\tag{43}$$

where

$$l = \frac{D}{8V_1} \quad \text{or} \quad l = \frac{1}{7.1\sqrt{w_0 V_1}}.\tag{44}$$

Here, as before, the parameter n is determined from the curve in Figure 4 using the value of v_M, but in the given case the expression for v_M is transformed to the form

$$v_M = 1.3\,\frac{w_0 D}{H}.\tag{45}$$

The parameter A retains the values indicated above, but changes its dimension by $\text{m}^{1/3}$. By using the value of v_M obtained from (45), from Figure 5 we find the value of d, and as a result determine x_M. In the case of cold discharges, v_M is frequently small. For these values of v_M the relationship $v_M \approx v_M$ holds according to (19), but v_M is calculated from (45).

3. Propagation of a heavy polydisperse contaminant

The basic principles for calculating the dispersion of heavy monodisperse particles in the atmosphere have been described in /6, 8/, and /12/. The initial formulas for calculating the concentrations of a heavy contaminant were established by integrating the equation of turbulent diffusion (3) when $w \neq 0$.

In general, the following formula is derived from the formulas given in /12/ for the relationship between the ground values of the concentrations of the heavy and light contaminants q_w and q at a distance x from a source of height H:

$$q_w = q\chi\left(\frac{w}{k_1},\ \frac{k_1 x}{u_1},\ H\right). \tag{46}$$

The function χ is determined according to the numerical solution of the problem.

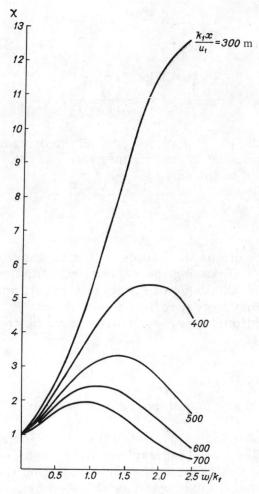

FIGURE 9. χ as a function of parameters $\frac{k_1 x}{u_1}$ and $\frac{w}{k_1}$.

The differences in the concentrations of the light and heavy contaminants for given values of H and $\frac{k_1 x}{u_1}$ are determined by the
dimensionless parameter $\frac{w}{k_1}$. This parameter characterizes the
influence of the gravitational precipitation in the process of turbulent diffusion of the contaminant. At a given value of w, the part
played by the precipitation will depend on the intensity of turbulence.
For strong turbulence, such as exists in developed convection, the
influence of differences in the rate of precipitation of the contaminant w decreases, and is noticeable mainly at sufficiently large
values of w and x. In this case for small values of w (lower than
3 cm/sec) $\chi = 1$.

In /8/ are given some results of the calculation of χ for monodisperse particles with dimensionless rates of precipitation $\frac{w}{k_1}$
varying between 0.25 and 1.25 for source heights H between 50 and
250 m, and up to effective distances $\frac{k_1 x}{u_1} = 750$ m.

We recently analyzed the results of more accurate calculations
of the distribution of the ground concentrations of a heavy contaminant, conducted with a finer mesh in the numerical solution of the
problem.* Such calculations were possible by using the M-220
computer, which is more than two orders of magnitude more rapid
than the computers of the Ural-1 series, and have four times the
core memory. Not only was the accuracy of the calculations
increased, but a wider range of variation in the parameters was
investigated: $\frac{w}{k_1} = 0 - 5.0$; $H = 20 - 500$ m, $\frac{k_1 x}{u_1} < 1000$ m.

In Figure 9 the results of calculations of χ for a discharge height
of 400 m are given in a form convenient for use, and based on
dimensional considerations. It is possible to draw a number of
general conclusions from this figure on the laws of diffusion of a
heavy monodisperse contaminant. It follows, in particular, that
near the source the concentrations of the heavy contaminant can
considerably exceed those of the light contaminant.

In the given case the concentration of the light contaminant is
maximum at an effective distance $\frac{k_1 x}{u_1} \approx 600$ m. The magnitude χ
reaches its maximum value, near to two, when $\frac{w}{k_1} \approx 1$.

* The computations were carried out with the participation of Gracheva and Kiselev.

At sufficiently large distances for a given $\frac{w}{k_1}$, the value of χ becomes less than one, as a result of the partial fall out of the contaminant onto the underlying surface.

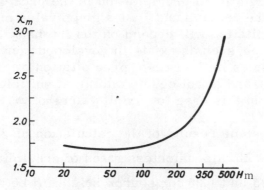

FIGURE 10. $\chi_m = \dfrac{q_{mw}}{q_m}$ as a function of the

source height H at $\dfrac{w}{k_1} = 1$.

It is interesting to note that according to the formulas given in /6/ the value of χ is the same for both averaged and nonaveraged concentrations, and is therefore independent of the sampling period.

Computer calculations for a monodisperse contaminant confirmed the analytical estimates made in /8/ and /12/. According to those, when $\frac{w}{k_1} = $ const, the parameter $\chi_m = \frac{q_{mw}}{q_m}$ is practically independent of the source height for sources located in the ground layer of the atmosphere, while for higher sources it increases rather rapidly with height. In Figure 10 we plotted the numerically obtained dependence of χ_m on H for $\frac{w}{k_1} = 1$.

When calculating the dispersion of real discharges of dust (sols), into the atmosphere, it is nearly always necessary to take into account their polydispersity, and also the density of the dust particles, ρ_d, on which the rate of their precipitation also depends. The rate of precipitation of dust in the atmosphere is usually determined by means of Stokes' formula, which can be written in the form

$$w = 1.3 \cdot 10^6 \, \rho_d r_d^2,$$

where w is in cm/sec, and the density (unit weight) of the particles ρ_d and their effective radius r_d are in g/cm^3 and cm, respectively.

22048

The investigations of the Dzerzhinskii All-Union Thermotechnical Institute, the Leningrad branch of the Gas Purification Scientific Research Institute, and the YuzhORGRES* concern, conducted in part during the testing of purification equipment, showed that the dispersive characteristics of the dust discharged into the atmosphere vary considerably as a function of the quality of the raw materials, the mode of operation of the technical equipment, and the purification devices. They are also dependent on the type of dust removal devices, their efficiency, and other factors. Unfortunately, the part played by the different factors has not been studied sufficiently, and there are examples for some industries only.

At the same time, for practical purposes we can restrict ourselves to a first approximation to typical values of these characteristics based on the data presented by these organizations. According to these data, ρ_d lies between 1 and $2-2.5 \text{ g/cm}^3$ for sols formed during coal combustion and for dust from apatite, nepheline, and many other industries. The dust discharges from ferrous and nonferrous metallurgy are frequently characterized by ρ_d greater than $3-4 \text{ g/m}^3$. The range of variation in r_d is between 0 and 50μ, and is even greater at times.

Instead of taking the density and size distribution of the particles separately, it seems more logical to consider their complex characteristic $P(w)$, the weight function of distribution of particles with different precipitation rates w_i. In fact, the functions given above explicitly contain the parameter w_i. Besides, in the case of efficient dust purification, this distribution must be more conservative than the size distribution of the particles, since for dust which has passed through dynamic purification devices there is a marked correlation between the size distribution of the particles, their density, and the dust-removal efficiency. Heavy particles are usually easier to eliminate, and as a result, for a given purification efficiency the increase in the unit weight of the dust is accompanied by a decrease in the maximum size of the particles which are most important in the size distribution of the cleaned dust. The maximum rate of precipitation changes less. For a $90-95\%$ efficiency, the maximum values of w_i which have an appreciable effect on the size distribution do not usually exceed 50 cm/sec. For a mean ρ_d between 1 and $2-2.5 \text{ g/cm}^3$, and an efficiency higher than 90%, we can consider that the weight of particles with w_i up to 5 cm/sec is $40-50\%$ of the total weight of the dust, with w_i from 5 to 25 cm/sec about $30-40\%$, and with higher rates of precipitation about 20%.

* [Southern State Trust for the Organization and Rationalization of Regional Electric Power Plants and Networks.]

In the calculations the standard distribution $P(w)$ is generally split into n equal intervals, to each of which corresponds a mean precipitation rate w_i, and at a fixed distance $\frac{k_1 x}{u_1}$ a value $\chi\left(\frac{w_i}{k_1}\right)$. The total value of the concentration is $q_w = \sum\limits_{i=1}^{n} q_{wi}$. Thus

$$q_w = q \sum_{i=1}^{n} P(w_i)\, \chi\left(\frac{w_i}{k_1}\right). \tag{47}$$

The calculations yield the maximum value of the total concentration q_{mw} and the distance x_{mw} at which it is reached. We can write

$$q_{mw} = q_m F, \tag{48}$$

$$x_{mw} = x_m \varkappa. \tag{49}$$

It is clear that $F>1$, $\varkappa<1$. It follows from (46) that in the general case F and $\varkappa$ can depend on the source height.

The values of F and $\varkappa$ were calculated by this procedure for $n=10$ and a number of characteristic dust distributions. It follows that the polydispersity of the dust leads to a spreading of the position of the maxima of the different fractions along the jet axis, and therefore leads to a decrease in the maximum of the total concentration.

The calculations showed that F increases somewhat with the source height. For the distribution indicated above, corresponding to a rather highly efficient ($\sim$95%) dust purification, F is equal to 1.5—2.0 up to effective source heights of 300—400 m and is somewhat larger for heights of 500 m.

From the calculations it also follows that in (48) we can take q_m with approximately the same exponent β_1 as in formula (8). Therefore, to a first approximation, at a mean value of F for a heavy polydisperse contaminant we can use the same formula for the dangerous wind velocity as for the gases.

From (48) for the maximum dust concentration $c_{M.d}$ under unfavorable meteorological conditions we obtain

$$c_{M.d} = c_M F, \tag{50}$$

where c_M is the maximum concentration of the light contaminant under unfavorable conditions.

For light contaminants $F=1$. We can take F as approximately equal to one, even for dust with a rate of precipitation not exceeding 2—3 cm/sec.

For the distance $x_{M.d}$ at which the maximum concentration of polydisperse dust $c_{M.d}$ is attained, from (49) we obtain

$$x_{M.d} = x_M \varkappa. \qquad (51)$$

The value of x_M is determined from the formula (39), while $\varkappa$ is in the general case a function of the source height (Table 2).

TABLE 2.

H(m)	100	150–200	250–300	350–400	450–500
$\varkappa$	0.75	0.70	0.65	0.60	0.55

We can take, on the average, $\varkappa = 0.7$.

It follows from formula (46) that a similar formula can also be written for the distance x_{Mud}, at which the maximum concentration of the dust c_{Mud} is reached at a wind velocity u different from u_M:

$$x_{Mud} = x_{Mu} \varkappa. \qquad (52)$$

Here $\varkappa$ is also determined from Table 2.

The concentration c_d on the jet axis is determined from the formula

$$c_d = c_{M.d} s_1 \left(\frac{x}{x_{M.d}} \right), \qquad (53)$$

and

$$c_{ud} = c_{Mud} s_1 \left(\frac{x}{x_{Mud}} \right). \qquad (54)$$

The dimensionless factor s_1 is determined from the curve in Figure 7, as for a light contaminant. We have made the previously used computing procedure of the axial concentrations more accurate. In the older methods, the difference btween x_M and $x_{M.d}$ was not considered and for dust with $\frac{x}{x_{M.d}}$ greater than 2, the value of s was calculated from a special graph.

Other types of distribution $P(w)$ were also considered, taking into account the variants corresponding to the maximum pollution of the ground air layer.

The value of F increases when the purification efficiency drops. For efficiencies of 90% and above, it is possible to take F as approximately 2 in design calculations. For efficiencies between 75 and 90%, the value of F must be increased roughly 1.25 times, and for less effective or nonexistent purifications 1.5 times. Since the calculations are somewhat tentative, the values of F obtained seem sufficient for practical purposes.

If under dangerous conditions the height of the source exceeds 350—400 m, allowing for the initial ascent, we must take into account the increase in F with increase in H.

For a source height of about 500 m, F increases by roughly 10%, which is equivalent to some decrease in β_1 in (8) when $F=$const.

In this case the parameter u_M is calculated as in the preceding section from the equation $\frac{\partial q_{mw}}{\partial u}=0$. It is easily seen that the dangerous wind velocity u_M then decreases.

4. Calculation of the field of concentrations from many sources

The transition to the calculation of the total pollution of the atmosphere from a group of sources can be based on the results described earlier in /12, 24/, and /32/. In particular, in the case of N closely grouped stacks with identical parameters, it is convenient to conduct the calculation by the earlier formulas (37), (43), with the substitution

$$V_1 = \frac{V}{N},\qquad (55)$$

where $V = \frac{\pi D^2}{4} w_0 N$ is the total volume of discharged gases. In the transformed formulas, M is understood to be the total discharge of harmful substances through all N stacks.

For sources with different dangerous wind velocities it was found that the maximum concentration is observed at a wind velocity roughly equal to the weighted mean dangerous velocity

$$u_{Mc} = \frac{\sum\limits_{i=1}^{N} c_{M.i}\, u_{M.i}}{\sum\limits_{i=1}^{N} c_{M.i}},\qquad (56)$$

where $c_{M.i}$ and $u_{M.i}$ are the values of the maximum concentration
and the dangerous wind velocity for the i-th source. Some other
results as well, obtained earlier for a relatively small number of
sources /12/, are still valuable.

The extension of the computing procedure to sources with a
wider range of variation in the discharge parameters poses new
problems with regard to the determination of the concentration field
from a group of sources. These problems are mainly due to the
necessity of allowing for a large number of varied sources.

When estimating the air pollution from sources dispersed over
the territory, considerable difficulties arise in the examination of
the concentration field for different wind directions, if the location
of the sources and the corresponding location and magnitude of the
maximum concentration vary appreciably. The formal approach
to the solution of this problem consists in calculating the concentration field at the nodes of some sufficiently dense mesh of points.
Besides, such calculations must be conducted for all possible wind
directions. The sufficiency of the mesh points and of the frequency
of sampling the wind direction can be estimated only on the basis of
a numerical experiment, which is very labor-consuming even when
using a computer, and consists in decreasing the mesh spacing, and
sampling a large number of wind directions.

We developed a more efficient approach, leading to a considerable saving in computer time. Its basic principles consist in
the following. The wind directions for which the calculations are
to be conducted must coincide with the lines connecting the most
powerful spaced sources; if several close wind directions are
obtained as a result, they are combined into one. Next, for each
given wind direction the positions of all the points are determined at
which maximum concentrations from each individual source are
observed. The total concentration field is now calculated, not at the
nodes of a regular mesh, but for points of computed maxima from
the individual sources. The analysis and the results of numerous
computer calculations show that the position of the resultant maximum cannot appreciably differ from one of these points; the errors
in the maximum concentration caused by this difference do not
generally exceed 5—10% /24/. In the given case, we can speak of
dangerous wind directions, corresponding to c_M.

In some cases, when the dangerous wind velocities $u_{M\,i}$ for the
different sources differ considerably from $u_{M\,c}$ calculated from (56),
the computer calculations are performed not only for $u=u_{M\,c}$ but also
for other wind velocities, usually equal to $u_{M\,i}$. It is thus possible
to make the calculations more accurate in the vicinities of sources
with $u_{M\,i}$ differing from $u_{M\,c}$.

This procedure was programmed at the Main Geophysical Observatory for the M-220 computer. The calculations showed that this program reduces computer running time by more than one order of magnitude, compared to the formal method of regular calculations. With the program it is possible to calculate c_M and the field of maximum possible concentrations for several hundred sources. Calculations have been conducted so far for a large city, and for different districts, taking into account up to 400 stacks. Some questions on the calculation of air pollution from many sources on a computer by the above procedure are discussed in detail in /24/.

An example of plants with numerous sources are rayon plants. Hydrogen sulfide and carbon disulfide are discharged in them mainly through 2—3 high stacks; there are also several hundred air shafts, chimney hoods, exhausts, and other sources mounted on the roof of the industrial premises. In such plants the high and low sources must be treated separately. The calculation must be conducted separately for the industrial area itself and for areas beyond it, such as residential quarters, which frequently lie far from the large industrial enterprises. The pollution of the air basin of the plant area can be approximately calculated by means of the computer program described in /24/.

The concentration produced by a set of small sources at a considerable distance from them can be calculated replacing the set by one or several equivalent sources. It follows from formulas (41) and (42) and the curves in Figures 7 and 8 that the highest resulting errors are caused by the arbitrary shift of the sources along the y axis. Therefore, for a given wind velocity from Figure 8, for the coefficient s_2 in formula (42), we can find the distance x_p from which such a reduction of the sources to the center of their area is admissible. If the relative error in the calculation of the total concentration is to be less than δ, then the inequality $s_2 > 1 - \delta$ must be true for all the sources. From the value of $s_2 = 1 - \delta$, and from

Figure 8, we determine the value of $\eta = \left(\dfrac{|y|}{x} \right)_{s_2 = 1 - \delta}$. If the size of

the given area in the y direction is equal to a_1, then $x_p > \dfrac{a_1}{2\eta}$. Thus,

for a typical rayon plant at $\delta = 0.1$ $x_p = 1 - 2$ km.

At distances greater than x_p it is possible in practice to neglect some differences in the height and other discharge parameters of the low sources. In fact, it was found in /3/ that at large distances from low sources the part played by the discharge height is small. The comparative calculations conducted by the approximate and exact procedures agreed satisfactorily /1/.

The concentrations from a set of small sources at a large distance from them obtained by this approximate method can be used as background concentrations for the determination of air pollution caused by organized discharges from high stacks.

It is advisable to conduct similar calculations for groups of small sources during the preliminary design stage, when sufficiently detailed information is not yet available on the location of the sources and only the gross discharge of harmful substances is known.   Such an approach considerably simplifies the tentative estimates of the possibility of building the plant in the given region, and of the admissible discharge of different contaminants into the atmosphere.

5. Experimental corroboration.
Conclusions and recommendations

Besides the results described on the extension of the computing procedure to the case of cold discharges, polydisperse dust, and a large number of sources, in the practical utilization of the theory of atmospheric diffusion many previously obtained conclusions on anomalously dangerous conditions of air pollution in the case of high inversions /7, 9/, calm weather /9, 10, 14/, mists /11/, unhomo-geneous relief /9, 13/ can be utilized, and also the calculation of the health-protective zone /12/, etc.

The results of the theoretical investigations were confirmed by comprehensive empirical data.   The results testing the method of calculation of the dispersion of hot discharges in the atmosphere, which is a particular case of the procedure described here, have already been published.   The theoretical conclusions were con-firmed by the observations performed during the expeditions of the Main Geophysical Observatory in the region of several powerful TPP with stack height between 120 and 180 m, in the sanitary in-vestigations of a large number of TPP with stack height between 20—30 and 150—180 m /6, 12, 22, 30/, and also of the Chelyabinsk metallurgical and the Chimkent lead plants /12/.

In 1970, the Main Geophysical Observatory organized several expeditions to study the air pollution round the powerful thermo-electric power plant in the Krivoi Rog district, with 250 m high stacks /2/.

It is interesting to note that while formula (1) for calculating the maximum concentration c_M as a function of the stack height H

and other parameters was developed for H less than 120 m, it was found to be completely applicable to the new 250 m stacks.

Some experimental studies of TPP areas have been published recently outside the USSR (/34/ and others); their results also agree with those of the calculation of the maximum concentration by our formulas.

For relatively cold discharges the first results on the testing of the computing procedure were obtained in measurements of the concentrations of nitrogen oxides under the jet of the Nevskii chemical plant /23/.

Some principles of the calculation of the dispersion of a heavy contaminant in the atmosphere were checked, and were satisfactorily confirmed by the results of the discharge of fluorescent dust from the 300 m high meteorological tower of the Institute of Experimental Meteorology /17/, and of the spraying of aerosols from an aircraft flying perpendicular to the wind at heights between 100 and 400 m /27/.

In this collection the results of four expeditions are described. The dispersion of cold or slightly heated exhaust discharges of hydrogen sulfide and carbon disulfide from rayon plants /18/ into the atmosphere was studied. The expeditions were conducted under different physicogeographical conditions: two in Povolzh'e, one in Ukraine, and one in Siberia. The program of the expeditions included measurement of the amount of hydrogen sulfide and carbon disulfide discharged into the atmosphere, the discharge velocity of the gases into the atmosphere, their temperature, etc. Simultaneously, the field of concentrations of these gases was determined up to distances of 8—15 km from the plants, and meteorological observations were carried out.

The data of the expeditions were used not only for comparison with results of the calculations of ground concentrations about the source, but also for checking and improving the design data with reference to the discharge parameters of contaminants into the atmosphere. The expeditions found some discharge sources which had previously been neglected when plants were designed. The results indicate agreement between the theory and the experiment for very different characteristics in both the cold and hot seasons of the year. In particular, the values of the maximum concentration of hydrogen sulfide and carbon disulfide differed from the calculated values within the limits of the accuracy of the measurements. The values of the calculated weighted mean dangerous wind velocities were about 1 m/sec in summer and 2—3 m/sec in winter. Such were also the wind speeds for which the measured concentrations were maximum. The concentrations are maximum in the regions immediately adjoining the plants; this follows from the calculations and the experimental results.

From the observed agreement between experiment and theory, for practical utilization we can recommend the results obtained in this paper for improving and extending the methods already suggested for calculating the dispersion of contaminants in the atmosphere /19, 32/. These methods ensure sufficient purity of the air basin at a minimum cost.

In the first place, it is expedient to decrease the number of stacks, as already indicated in /6/ and /12/. It follows from (37) and (55) that in the case of closely arranged stacks with identical discharge parameters

$$c_M = \frac{AMmnF}{H^2} \sqrt[3]{\frac{N}{V \Delta T}}. \tag{57}$$

At a sufficiently large ΔT, more exactly, when $v_M > 2$ m/sec, $c_M \sim \sqrt[3]{N}$, since $n = 1$, and m varies little as a function of N. This law is of great practical importance. It follows that for given discharge characteristics the ground concentrations will be somewhat lower when the number of stacks is reduced. This can result in a considerable saving. In fact, the cost of one single 250 m high stack for a typical 2,400 MW TPP is about 2 million rubles.

While the concentration c_M decreases with decrease in N, the volume V of discharged gases remains constant. If the drift velocity w_0 does not change either, the reduction in the number of stacks must be accompanied by an increase in the diameter D of their opening. The dependence of c_M on N can be explained by the characteristics of the initial ascent ΔH. It follows from (9) that when D is increased and the other parameters are kept constant, ΔH increases. This is also true at the dangerous wind velocity u_M. In particular, at low values of f, when $u_M \approx v_M$, the corresponding expression for ΔH is in the form

$$\Delta H|_{u=u_M} \approx H(0.17 + 0.3\sqrt[3]{f}). \tag{58}$$

We can therefore state that the initial ascent from stacks of larger diameter is higher, and as a result the ground concentration is lower.

Another fact to be considered is also associated with the analysis of the character of the dependence of c_M on the output parameters. It follows from (57) that c_M is affected most by the stack height H, since $c_M \sim \frac{1}{H^2}$. The next factor in importance is the magnitude of

the discharge of harmful substances M, since according to (57)
$c_M \sim M$. The value of c_M depends much less on other discharge
parameters, in particular on V, w_0, D, ΔT.

The functional dependence of c_M on the output parameters varies
appreciably with the transition to relatively cold discharges, or
more exactly to sources with large f and small v_M. This is shown
most clearly in the limiting case $(\Delta T = 0)$, for which

$$c_M = \frac{AMnF}{H^{4/3}} l. \tag{59}$$

If we also take into account that in this case $n \sim H^{1/3}$, it follows
from (59) that $c_M \sim \frac{1}{H}$. Thus for cold discharges the source height has a
much smaller influence on the ground concentration than for hot
discharges.

For N stacks, from (43), (44) and (55) we see that in (59) we
must assume

$$l = \frac{1}{7.1\sqrt{w_0}} V \sqrt{\frac{N}{V}}. \tag{60}$$

Since the value of n is little dependent on N, we have $c_M \sim \sqrt{N}$.
Therefore, for cold discharges the dependence of c_M on N is likewise
greater than for hot discharges. This means that a reduction in the
number of stacks will be even more effective in the case of cold
exhausts.

The theoretical conclusion /19, 32/ on the effectiveness of
reducing the number of stacks is being increasingly reflected in
contemporary designs. The number of stacks in large power plants
has been reduced considerably, and the same trend is observed in
oil-refining plants, in many large metallurgical plants, etc.

At the same time, it is not always possible to concentrate the
discharges in a single stack. This is partly due to the necessity of
periodically overhauling stacks damaged by corrosion, high temperatures, etc. If there is only one stack, the production process will
be greatly disturbed. Therefore, there is a current trend to use
multi-shaft stacks, in particular in powerful TPP /31/. These
stacks consist of several pipes on one supporting construction, and
enclosed at times in a common casing. In this case the relationships given above between c_M and N hold only over a certain range
of distances between the pipes, and have both upper and lower limits.
The upper limit coincides with the distance $0.5H—H$. For such a
distance the difference between the resultant maximum and the sum

of the maxima of the concentrations from the individual sources is sufficiently small. This difference increases appreciably with further increase in the distance between the sources.

When the pipes are very close, there is also a lower limit to the possibility of using the given relationships between c_M and N. Its evaluation is of considerable interest in view of the above trend toward multi-shaft stacks.

It can be assumed that when the pipes are in contact, their smoke jets practically coalesce. In this limiting case the correction factor in c_M for the number of stacks N is equal to one, i. e., the concentration will be the same as for one stack.

When the distances between the stacks are equal to several stack diameters, then into the calculations conducted for $N = 1$ we must introduce a correction factor larger than one but smaller than $\sqrt[3]{N}$ or $\sqrt{N}$ for formulas (57) and (59), respectively. It is possible to calculate tentatively the minimum distance at which formula (57) can be used from the condition that the smoke jets come into contact at a height $\Delta H|_{u=u_M}$ (58) above the source.

If we assume for this calculation that the angle of spread of the jet is approximately 0.2, this distance will be equal to $0.1 \Delta H|_{u=u_M}$. In the case of a multi-shaft stack with N sections situated at a distance from one another of less than $0.1 \Delta H|_{u=u_M}$, it is possible to calculate the concentration c_M for $N = 1$, and to multiply it by some factor found by interpolation between 1 and $\sqrt[3]{N}$ or $\sqrt{N}$, depending on the given case. The maximum distance between the sections must be used in the interpolation. The results of experiments on the modeling of smoke jets from a number of closely located stacks are also useful for such an evaluation.

Finally, we can conclude that from the results described above the limits of applicability of the methods for computing the dispersion of industrial exhausts into the atmosphere can be appreciably extended.

Bibliography

1. Baikov, B. K. et al. Proverka metodiki rascheta rasseivaniya v atmosfere kholodnykh vybrosov na materialakh obsledovaniya predpriyatii iskusstvennogo volokna (Test of the Method for Computing the Dispersion in the Atmosphere of Cold Discharges on Materials from the Study of Rayon Plants). See this Collection.

2. B e l u g i n a, V. A. et al.   Issledovanie rasprostraneniya vred-
    nykh primesei v raione vysokikh istochnikov vybrosa
    (Study of the Propagation of Contaminants in the
    Region of High Discharge Sources). See this Collection.
3. B e r l y a n d, M. E.   K teorii turbulentnoi diffuzii (The Theory
    of Turbulent Diffusion). − Trudy GGO, No. 138, pp. 31−37.
    1963.
4. B e r l y a n d, M. E.   Ob opasnykh usloviyakh zagryazneniya
    atmosfery promyshlennymi vybrosami (Dangerous Con-
    ditions of Pollution of the Atmosphere by Industrial
    Wastes). − Trudy GGO, No. 185, pp. 15−25. 1965.
5. B e r l y a n d, M. E.   Meteorologicheskie problemy obespeche-
    niya chistoty atmosfery (Meteorological Problems of the
    Maintenance of a Pure Atmosphere). − Meteorologiya i
    Gidrologiya, No. 11. 1967.
6. B e r l y a n d, M. E., E. L. G e n i k h o v i c h, and R. I. O n i k u l.
    O raschete zagryazneniya atmosfery vybrosami iz dymo-
    vykh trub elektrostantsii (Calculation of Atmospheric
    Pollution by Discharges from the Stacks of Power
    Plants). − Trudy GGO, No. 158, pp. 3−21. 1964.
7. B e r l y a n d, M. E. et al.   Chislennoe issledovanie atmosfernoi
    diffuzii pri normal'nykh i anomal'nykh usloviyakh strati-
    fikatsii (Numerical Investigation of Atmospheric Diffusion
    under Normal and Abnormal Stratification Conditions). −
    Trudy GGO, No. 158, pp. 22−32. 1964.
8. B e r l y a n d, M. E. et al.   Osobennosti diffuzii tyazheloi pri-
    mesi v atmosfere (Diffusion of a Heavy Contaminant into
    the Atmosphere). − Trudy GGO, No. 158, pp. 33−40. 1964.
9. B e r l y a n d, M. E., E. L. G e n i k h o v i c h, and V. K. D e m ' y a n o -
    v i c h.   Nekotorye aktual'nye voprosy issledovaniya
    atmosfernoi diffuzii (Some Topical Problems of the Study
    of Atmospheric Diffusion). − Trudy GGO, No. 172, pp. 3−22.
    1965.
10. B e r l y a n d, M. E. et al.   Vliyanie vertikal'nogo rasprostrane-
    niya temperatury i skorosti vetra na atmosfernuyu diffuziyu
    radioaktivnykh primesei (Influence of the Vertical Propa-
    gation of the Temperature and the Wind Velocity on the
    Atmospheric Diffusion of Radioactive Contaminants).
    Moskva, Atomizdat. 1965.
11. B e r l y a n d, M. E., R. I. O n i k u l, and G. V. R y a b o v a.   K teorii
    atmosfernoi diffuzii v usloviyakh tumana (The Theory of
    Atmospheric Diffusion under Fog Conditions). − Trudy
    GGO, No. 207, pp. 3−13. 1968.

12. Berlyand, M. E. and R. I. Onikul. Fizicheskie osnovy rascheta rasseivaniya v atmosfere promyshlennykh vybrosov (Physical Basis of the Calculation of the Dispersion of Industrial Discharges in the Atmosphere). — Trudy GGO, No. 234, pp. 3—27. 1968.

13. Berlyand, M. E., E. L. Genikhovich, and O. I. Kurenbin. Vliyanie rel'efa na rasprostranenie primesi ot istochnika (Influence of Relief on the Propagation of a Contaminant from a Source). — Trudy GGO, No. 234, pp. 28—44. 1968.

14. Berlyand, M. E. and O. I. Kurenbin. Ob atmosfernoi diffuzii primesei pri shtile (Atmospheric Diffusion of Contaminants in Calm Weather). — Trudy GGO, No. 238. 1969.

15. Budyko, M. I. Isparenie v estestvennykh usloviyakh (Evaporation under Natural Conditions). Leningrad, Gidrometeoizdat. 1968.

16. Budyko, M. I. and L. S. Gandin. Opredelenie turbulentnogo obmena mezhdu okeanom i atmosferoi (Determination of the Turbulent Exchange between the Ocean and the Atmosphere). — Meteorologiya i Gidrologiya, No. 11. 1966.

17. Byzova, N. L. and R. I. Onikul. Analiz polya kontsentratsii tyazheloi primesi po dannym opytov na 300-metrovoi meteorologicheskoi machte (Analysis of the Concentration Field of a Heavy Contaminant on the Basis of Measurements Conducted on a 300-m High Meteorological Tower). — Trudy GGO, No. 172, pp. 35—41. 1965.

18. Vdovin, B. I. et al. Eksperimental'nye issledovaniya rasseivaniya v atmosfere kholodnykh ventilyatsionnykh vybrosov predpriyatii iskusstvennogo volokna (Experimental Studies of the Dispersion in the Atmosphere of Cold Ventilation Discharges of Rayon Plants). See this Collection.

19. Vremennaya metodika raschetov rasseivaniya v atmosfere vybrosov (zoly i sernistykh gazov) iz dymovykh trub elektrostantsii (Provisional Method for Calculating the Dispersion in the Atmosphere of Discharges (Sols and Sulfur Dioxide) from the Stacks of Power Plants). — Trudy GGO, No. 172, pp. 205—212. 1965.

20. Genikhovich, E. L. and V. P. Gracheva. Analiz dispersii gorizontal'nykh kolebanii napravleniya vetra (Analysis of the Variance in the Horizontal Fluctuations of the Wind Direction). — Trudy GGO, No. 172, pp. 42—47. 1965.

21. Gil'denskiol'd, R. S. and B. V. Rikhter. Gigienicheskaya
    otsenka deistvuyushchei metodiki rascheta rasseivaniya
    vybrosov GRES v atmosfernom vozdukhe (Health Evaluation
    of the Contemporary Method for Calculating the Dispersion
    of Discharges from the State Regional Electric Power Plant
    (GRES) in the Atmosphere). — In: Voprosy gigieny, plani-
    rovki i sanitarnoi okhrany atmosfernogo vozdukha. Moskva.
    1966.
22. Gil'denskiol'd, R. S. et al. Rezul'taty eksperimental'nykh
    issledovanii zagryaznenii atmosfery v raione Moldavskoi
    GRES (Results of Experimental Investigations on the
    Pollution of the Atmosphere in the Region of the Moldavian
    State Regional Electric Power Plant). — Trudy GGO, No. 207,
    pp. 65—68. 1968.
23. Goroshko, B. B. et al. Rezul'taty nablyudenii za zagryaz-
    neniem atmosfery okislami azota ot khimicheskogo zavoda
    (Results of Observations on Atmospheric Pollution by
    Nitrogen Oxides from a Chemical Plant). — Trudy GGO,
    No. 185. 1966.
24. Gracheva, I. G. et al. K raschetu zagryazneniya atmosfery
    ot mnogikh istochnikov (Calculation of the Pollution of the
    Atmosphere by Many Sources). — Trudy GGO, No. 238.
    1969.
25. Gracheva, V. P. and V. P. Lozhkina. Ob ustoichivosti
    napravleniya vetra v prizemnom sloe atmosfery (The
    Stability of the Wind Direction in the Ground Layer of the
    Atmosphere). — Trudy GGO, No. 158, pp. 41—45. 1964.
26. Gracheva, V. P. K kharakteristike termicheskoi ustoichivosti
    v prizemnom sloe vozdukha (On the Characteristics of the
    Thermal Stability in the Ground Layer of the Atmosphere).—
    Trudy GGO, No. 238. 1969.
27. Dunskii, V. F., I. S. Nezdyurova, and R. I. Onikul. O
    raschete rasseivaniya osedayushchei primesi ot lineinogo
    istochnika v pogranichnom sloe atmosfery (On the Cal-
    culation of the Diffusion of a Precipitating Contaminant
    from a Linear Source in the Boundary Layer of the
    Atmosphere). — Trudy GGO, No. 207, pp. 28—37. 1968.
28. Kazanskii, A. B. and A. S. Monin. O turbulentnom rezhime
    v prizemnom sloe vozdukha pri neustoichivoi stratifikatsii
    (The Turbulent Regime in the Boundary Air Layer during
    Unstable Stratification). — Izvestiya AN SSSR, Seriya Geo-
    fizicheskaya, No. 6. 1958.

29. Monin, A. S. and A. M. Yaglom. Statisticheskaya gidro-
      mekhanika. Mekhanika turbulentnosti. Ch. 1 (Statistical
      Hydromechanics. Mechanics of Turbulence. Part 1).
      Moskva, "Nauka." 1965.
30. Onikul, R. I. et al. Rezul'taty analiza eksperimental'nykh
      dannykh, kharakterizuyushchikh raspredelenie atmosfer-
      nykh zagryaznenii vblizi teplovykh elektrostantsii (Results
      of the Analysis of Experimental Data on the Distribution of
      Atmospheric Contaminations Near Thermal Power Plants).—
      Trudy GGO, No. 172, pp. 23—34. 1965.
31. Rikhter, L. A. Gazovozdushnye trakty teplovykh elektro-
      stantsii (Gas-Air Channels of Thermal Power Plants).
      Moskva, "Energiya." 1969.
32. Ukazaniya po raschetu rasseivaniya v atmosfere vrednykh
      veshchestv (pyli i sernistogo gaza), soderzhashchikhsya
      v vybrosakh promyshlennykh predpriyatii. CH 369-67
      (Instructions Regarding the Calculation of the Dispersion
      in the Atmosphere of Contaminants (Dust and Sulfur Dio-
      xide) Contained in the Discharges of Industrial Enter-
      prises. CH 369-67). Leningrad, Gidrometeoizdat. 1967.
33. Berlyand, M. E., O. J. Kurenbin, and J. M. Zrajevsky.
      Theory of Tracer Distribution near Ground.— J. Geo-
      phys. Res., Vol. 75, No. 18. 1970.
34. Meteorology and Atomic Energy (D. H. Sladeed).— U. S. Atomic
      Energy Commission. 1968.

SIMULATION OF ATMOSPHERIC TURBULENT FLOWS ABOVE AN INHOMOGENEOUS UNDERLYING SURFACE

I. M. Zrazhevskii, V. V. Klingo

Introduction

The existing methods for calculating the dispersion of contaminants discharged into the atmosphere by industrial plants are
mainly applicable to individual sources located above a flat underlying surface /1/.

Lately, there has been increased interest in the propagation of
contaminants in streams with considerable gradients of the mean
velocity /2, 3/. Such gradients of the mean velocity appear when the
atmospheric stream flows past different types of inhomogeneities
of the ground surface, such as broken terrain or town buildings.

In view of the difficulty of studying atmospheric diffusion in a
complex relief by theoretical or field studies /12, 17/, simulation
of these processes seems a practical solution, and studies are being
conducted in this direction. In the wind tunnel of the Moscow
University Institute of Mechanics the mean velocity profiles and
some turbulent characteristics at flows past different models of
rugged terrain have been determined /1, 5/.

Simulation studies have also been carried out abroad /19—21, 24,
26/. These studies attempted to establish criteria of similitude for
turbulent flows, and to develop some practical suggestions on the
procedure for experimental studies.

The simulation of the atmospheric processes is fairly complex,
even in a simplified formulation, and is closely connected with
fundamental questions of the theory of turbulence. Besides, the
question of the criteria of similitude has not yet been solved satisfactorily.

In this paper we suggest a new approach to the determination of
similitude criteria for atmospheric turbulent flows, and at this
stage restrict our studies to neutral stratification of the atmosphere.
This approach is based on the description of turbulent flow by means
of Davydov's closed system of equations /8—10/. Some of this
author's ideas have already been used by Monin /13/ in his study
on the boundary layer of the atmosphere.

The given system of equations of turbulence is solved by means of certain hypotheses, which can be confirmed experimentally only. Accordingly, we use experimental data and some known theoretical principles to test the theoretical conclusions derived from Davydov's system of equations.

We shall follow Kirpichev /11/ in establishing the physical similitude of the above processes and the procedure for establishing criteria of similitude.

Initial system of equations

Davydov proposed a closed system of equations for the mean velocity, the second and third one-point moments of the velocity fluctuations, the dissipation and the dissipation flow of turbulent energy.

The main hypotheses used by Davydov in closing the system are based on the analogy between the dynamics of turbulent flow and the molecular-kinetic theory of an ideal gas, and on the expression of the fourth moments of the different fluctuating magnitudes through the second ones.

We used Davydov's system of equations for the following reasons.

1. The system of equations is closed (some authors, for example /24/, started from open systems of equations to describe turbulent flow. This is clearly unsuitable for establishing criteria of similitude, even from the point of view of the most general principles of the theory of similitude).

2. The system contains the fundamental turbulent characteristics, some of which can even at present be determined experimentally with sufficient accuracy, while others will probably be measured in the near future.

3. The system describes turbulence without any limitations whatsoever on any symmetry of flow of the process.

Finally, we believe that this system is the best of all closed systems so far suggested to describe turbulence. Also, as indicated in the introduction, some hypotheses based on its closure have already been applied to describe atmospheric turbulence /13/.

When turbulent diffusion is studied above an uneven underlying surface at neutral stratification, from Davidov's system we can find both the mean velocity field and the characteristics of turbulent diffusion in terms of the turbulence parameters mentioned.

We should note that the equations under consideration are non-stationary. They contain explicitly the dissipation of turbulent energy and do not include any sources of turbulence. Therefore, the solutions of these equations must decrease in time and describe degenerate turbulence. In the simulation we must deal with a stationary stream (we neglect turbulence degeneracy in the working section of the wind tunnel). It is therefore advisable to introduce external sources of energy or random forces distributed over the whole space into the equations. Random forces were similarly introduced in /27/ in the investigation on stationary turbulence.

We must also allow for the action of the Coriolis force on the atmosphere by introducing suitable terms into the Davydov equations. The initial system will then include the following equations: equation for the mean velocity

$$u_k \frac{\partial u_i}{\partial x_k} + \frac{\partial R_{ik}}{\partial x_k} + \frac{\partial P}{\partial x_i} + \overline{K}_i = \nu \frac{\partial^2 u_i}{\partial x_k^2}; \qquad (1)$$

equation of continuity

$$\frac{\partial u_k}{\partial x_k} = 0;$$

equation for the second moments

$$u_k \frac{\partial R_{ij}}{\partial x_k} + R_{ik} \frac{\partial u_j}{\partial x_k} + R_{jk} \frac{\partial u_i}{\partial x_k} + \frac{\partial S_{ijk}}{\partial x_k} + \frac{\beta}{R} Q \left(R_{ij} - \frac{1}{3} \delta_{ij} R \right) +$$

$$+ B_{ij} - \frac{1}{3} \delta_{ij} B_{kk} + \frac{2}{3} \delta_{ij} Q + \overline{K_i' v_j} + \overline{K_j' v_i} - F_{ij}^{(2)} = \nu \frac{\partial^2 R_{ij}}{\partial x_k^2}; \qquad (2)$$

equation for the third moments

$$u_m \frac{\partial S_{ijk}}{\partial x_m} + S_{ijl} \frac{\partial u_k}{\partial x_l} + S_{jkl} \frac{\partial u_i}{\partial x_l} + S_{ikl} \frac{\partial u_j}{\partial x_l} + R_{il} \frac{\partial R_{jk}}{\partial x_l} + R_{jl} \frac{\partial R_{ik}}{\partial x_l} +$$

$$+ R_{kl} \frac{\partial R_{ij}}{\partial x_l} + \beta_1 Q \frac{S_{ijk}}{R} + \overline{K_i' v_j v_k} + \overline{K_j' v_i v_k} + \overline{K_k' v_i v_j} - F_{ijk}^{(3)} = 0; \qquad (3)$$

equation for the dissipation of turbulent energy

$$u_k \frac{\partial Q}{\partial x_k} + \frac{\partial C_k}{\partial x_k} + \alpha \frac{Q}{R} R_{ik} \frac{\partial u_i}{\partial x_k} + 4 \frac{Q^2}{R} - F^{(4)} = \nu \frac{\partial^2 Q}{\partial x_k^2}; \qquad (4)$$

and, finally, equation for the dissipation flow of turbulent energy

$$u_k \frac{\partial C_i}{\partial x_k} + C_n \frac{\partial u_i}{\partial x_n} + R_{il} \frac{\partial Q}{\partial x_l} + \frac{2}{9} Q \frac{\partial R_{im}}{\partial x_m} + \frac{\beta_2 Q}{R} C_i + \overline{\nu K_i' \left(\frac{\partial v_k}{\partial x_l} \right)^2} -$$

$$- F_i^{(5)} = 0. \qquad (5)$$

The following symbols were used in equations (1)–(5): u — mean velocity; v — fluctuating velocity; P — mean pressure divided by the density of the medium;

$$R_{ij} = \overline{v_i v_j}; \qquad S_{ijk} = \overline{v_i v_j v_k};$$
$$R = \overline{v_i v_i}; \qquad Q = \overline{\nu \left(\frac{\partial v_i}{\partial x_k}\right)^2}, \tag{6}$$

where Q is the dissipation of turbulent energy (we neglect the other terms that enter in the accurate expression for Q);

$$C_i = \overline{\nu v_i \left(\frac{\partial v_k}{\partial x_l}\right)^2}, \tag{7}$$

where C_i is the dissipation flow of turbulent energy, ν is the kinematic viscosity of the medium;

$$K_i = f[(u_1 + v_1)\delta_{2i} - (u_2 + v_2)\delta_{1i}];$$
$$K_i' = f(v_1 \delta_{2i} - v_2 \delta_{1i}), \tag{8}$$

where f is the Coriolis parameter, equal to double the angular velocity of rotation of the earth,

$$\delta_{ij} = \begin{cases} 1 & \text{for } i = j, \\ 0 & \text{for } i \neq j; \end{cases}$$

β. α, β_1 β_2 are empirically determined numerical constants (their specific numerical values are not necessary for our purposes);

$$B_{ij} = \frac{1}{2} R_{ni} \frac{\partial u_n}{\partial x_j} \left(2 - \beta + 3\beta^2 \frac{R_{lm}^2}{R^2}\right);$$

$F^{(i)}$ are additional terms resulting from the introduction of a random force into the Navier-Stokes equations.

Summation is performed with reference to twice repeated subscripts. The bar indicates averaging. The X axis is in the direction of the mean wind, and the Z axis perpendicular to the ground surface.

The terms associated with the Coriolis force are obtained by replacing in Davydov's system

$$\frac{\partial u_i}{\partial t} \text{ by } \frac{\partial u_i}{\partial t} + f(u_1 \delta_{2i} - u_2 \delta_{1i}),$$

$$\frac{\partial v_i}{\partial t} \text{ by } \frac{\partial v_i}{\partial t} + f(v_1 \delta_{2i} - v_2 \delta_{1i}). \tag{9}$$

Next, in accordance with the procedure for finding similitude criteria /11/, we must represent the initial system (1)–(5) in a dimensionless form. We therefore formally notate all the magnitudes entering these equations in the form of products of the scale for the given magnitude by a dimensionless quantity. (We denote the scales by zero and the dimensionless quantities by a wavy bar:)

$$u_i = U_i^0 \tilde{u}_i; \quad Q = Q^0 \tilde{Q}; \quad x_i = x_i^0 \tilde{x}_i \quad R = R^0 \tilde{R};$$

$$R_{ik} = R_{ik}^0 \tilde{R}_{ik}; \quad C_i = C_i^0 \tilde{C}_i; \quad P = P^0 \tilde{P}; \quad f = F^0 \tilde{f};$$

$$S_{ijk} = S_{ijk}^0 \tilde{S}_{ijk}; \quad F^{(i)} = F^{(i)0} \tilde{F}^{(i)};$$

$$\nu = N^0 \tilde{\nu}. \tag{10}$$

The physical meaning of the scales introduced will be determined in the course of our examination of specific cases of simulation of flow past obstructions in the atmosphere.

By substituting expressions (10) in equations (1)–(5) and dividing by the coefficient of the corresponding term, we obtain a set of dimensionless complexes.

From (1) after division by the coefficient of $u_k \dfrac{\partial u_i}{\partial x_k}$, we obtain:

$$\frac{X_k^0 U_{1,2}^0 F^0}{U_k^0 U_i^0}; \quad \frac{R_{ik}^0}{U_k^0 U_i^0};$$

$$\frac{P^0 X_k^0}{X_i^0 U_k^0 U_i^0}; \quad \frac{\nu^0}{X_k^0 U_k^0}. \tag{11}$$

From the equation of continuity we obtain:

$$\frac{U_2^0 X_1^0}{X_2^0 U_1^0}; \quad \frac{U_3^0 X_1^0}{X_3^0 U_1^0}.$$

From (2) after division by the coefficient of $u_k \dfrac{\partial R_{ij}}{\partial x_k}$, we obtain:

$$\frac{F^0 R_{1,2;j}^0 X_k^0}{U_k^0 R_{ij}^0}; \quad \frac{F^0 R_{1,2;i}^0 X_k^0}{U_k^0 R_{ij}^0}; \quad \frac{R_{ik}^0 U_j^0}{R_{ij} U_k^0};$$

$$\frac{R_{jk}^0 U_i^0}{R_{ij}^0 U_k^0}; \quad \frac{S_{ijk}^0}{U_k^0 R_{ij}^0}; \quad \frac{Q^0 X_k^0}{R^0 U_k^0}; \quad \frac{Q^0 X_k^0}{R_{ij}^0 U_k^0};$$

$$\frac{R_{ni}^0 U_n^0 X_k^0}{R_{ij}^0 U_k^0 X_j^0}; \quad \frac{R_{ni}^0 U_n^0 X_k^0 R_{lm}^{0^2}}{R_{ij}^0 U_k^0 X_j^0 R^{0^2}};$$

$$\frac{R_{ni}^0 U_n^0 X_k^0}{R_{ij}^0 U_k^0 X_j^0}; \quad \frac{R_{ni}^0 U_n^0 X_k^0 R_{lm}^{0^2}}{R_{ij}^0 U_k^0 X_j^0 R^{0^2}}; \quad \frac{F^{0(2)} X_k^0}{R_{ij}^0 U_k^0}. \tag{12}$$

From (3) we obtain after division by the coefficient of $u_m \dfrac{\partial S_{ijk}}{\partial x_m}$:

$$\frac{F^0 S_{1,2;\,jk}\, X_m}{U_m^0\, S_{ijk}^0}\,; \quad \frac{F^0 S_{1,2;\,ik}^0\, X_m^0}{U_m^0\, S_{ijk}^0}\,; \quad \frac{F^0 S_{1,2;\,ij}^0\, X_m^0}{U_m^0\, S_{ijk}^0}\,; \quad \frac{S_{ijl}^0\, U_k^0 X_m^0}{S_{ijk}^0 U_m^0 X_i^0}\,;$$

$$\frac{S_{jkl}^0\, U_i^0 X_m^0}{S_{ijk}^0 U_m^0 X_l^0}\,; \quad \frac{S_{ikl}^0\, U_j^0 X_m^0}{S_{ijk}^0 U_m^0 X_l^0}\,; \quad \frac{R_{il}^0\, R_{jk}^0 X_m^0}{S_{ijk}^0 U_m^0 X_l^0}\,; \quad \frac{R_{jl}^0\, R_{ik}^0 X_m^0}{S_{ijk}^0 U_m^0 X_l^0}\,;$$

$$\frac{R_{kl}\, R_{ij}^0 X_m^0}{X_l^0 U_m^0 S_{ijk}^0}\,; \quad \frac{Q^0 X_m^0}{R^0 U_m^0}\,; \quad \frac{F^{0(3)}\, X_m^0}{U_m^0\, S_{ijk}^0}\,. \tag{13}$$

From (4) we obtain after division by the coefficient of $u_k \dfrac{\partial Q}{\partial x_k}$:

$$\frac{C_n^0\, X_k^0}{X_n^0 U_k^0 Q^0}\,; \quad \frac{R_{im}^0\, U_i^0 X_k^0}{R^0 X_m^0 U_k^0}\,; \quad \frac{Q^0 X_k^0}{R^0 U_k^0}\,; \quad \frac{F^{0(4)}\, X_k^0}{Q^0 U_k^0}\,. \tag{14}$$

From (5) we obtain after division by the coefficient of $u_k \dfrac{\partial C_i}{\partial x_k}$:

$$\frac{F^0 C_{1,2}^0\, X_k^0}{U_k^0\, C_i}\,; \quad \frac{C_n^0 U_i^0 X_k^0}{C_i^0 U_k^0 X_n^0}\,. \tag{15}$$

The correlations (11)–(15) give all the possible dimensionless combinations which may be criteria of similitude for the processes described by equations (1)–(5). To select criteria of similitude from these combinations, we must specify the problem by formulating suitable boundary conditions. Then, according to (11), only complexes from (11)–(15) which include parameters given on the boundary, and inner parameters of the system (such as viscosity, density, etc.) will be criteria of similitude. To these must be added dimensionless combinations of magnitudes given on the boundary and not included in (11)–(15), if there exist such parameters.

Simulation of obstructions in the atmosphere

We shall consider the dynamic effect of obstructions on an impinging turbulent stream. Not only a physical basis for the criteria of similitude found is important in simulation, but also the actual possibility of varying these criteria (in a suitable laboratory installation), and their simultaneous experimental determination. Therefore, the number of criteria of similitude must be small, and the measurement and variation in values given to these magnitudes must be easy and reliable.

Accordingly, we shall schematize the given process. Simulation in a wind tunnel is conducted without rotation, and therefore the Coriolis force is not explicitly considered. We shall compare the terms containing the Coriolis force parameter with the other terms of the equations.

From the condition

$$\frac{X_k^0 U_{1,2}^0 F^0}{U_k^0 U_i^0} \ll \frac{R_{ik}^0}{U_k^0 U_i^0}$$

we then obtain

$$\frac{F^0 X^0 U^0}{R^0} \ll 1, \tag{16}$$

or

$$F^0 \ll \frac{R^0}{X^0 U^0}. \tag{17}$$

This condition is sufficient for neglecting the influence of the Coriolis force in equation (1).

From the condition

$$\frac{F^0 R_{1,2;\,j}^0 X_k^0}{U_k^0 R_{ij}^0} \ll \frac{Q^0 X_k^0}{R^0 U_k^0}$$

we have

$$F^0 \ll \frac{Q^0}{R^0}. \tag{18}$$

This condition is sufficient for neglecting the influence of the Coriolis force in equation (2).

Finally, from the conditions:

$$\frac{F^0 S_{1,2;\,jk}^0 X_m^0}{U_m^0 S_{ijk}^0} \ll \frac{Q^0 X_m^0}{R^0 U_m^0} \quad \text{and} \quad \frac{F^0 C_{1,2}^0 X_k^0}{U_k^0 C_i^0} \ll \frac{Q^0 X_k^0}{R^0 U_k^0}$$

we once more obtain the inequality (18).

Thus, if inequalities (17) and (18) hold, the Coriolis force in the atmosphere can be neglected. Numerical assessments show that if we take the values characteristic of the atmosphere: $U^0 \sim 10^3$ cm/sec, $Q^0 \sim 1$ cm^2/sec, 3, $R^0 \sim 10^4$ cm^2/sec^2, then even for $X^0 = 10^5$ cm, (17) and (18) do not hold.

Due to technical difficulties associated with the rotation of the wind tunnel, it does not seem possible to create a boundary layer on the plate completely similar to the atmospheric (with rotation of the mean velocity in height). It is, however, known that the Coriolis force not only causes a rotation of the wind with height, but also leads to a sharp difference between the vertical distribution of the turbulence energy and its dissipation inside the boundary layer on the plate (without rotation of the system) and inside the atmospheric boundary layer. If in the wind tunnel we use the boundary layer on the plate (without rotation of the system), as model of the boundary layer of the atmosphere, the Coriolis force is not taken into account. If we simulate the actually existing vertical distribution of the turbulent characteristics, we indirectly allow for the influence of the Coriolis force on the formation of the impinging stream. Thus, the impinging stream can be simulated with an accuracy up to the rotation of the wind with height.

We should note that in /24/ and /19/, as well as in many other papers, it is assumed that the Coriolis force can be neglected in the study of flow past obstructions if the region studied does not exceed 150 km in the horizontal direction.

Next, when we study the flow past obstructions several hundred meters in height, we separate the impinging atmospheric stream into two regions (I and II), represented in Figure 1.

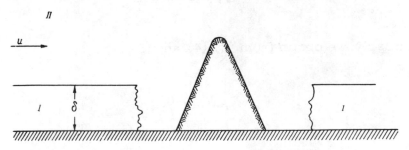

FIGURE 1. Schematic representation of the separation of the impinging atmospheric flow into regions I and II

The inner region I corresponds to the ground layer of the atmosphere in nature and to the boundary layer in the wind tunnel. This region is characterized by a logarithmic profile of the mean velocity, i. e., by a constant "dynamic velocity" v_*. Its height in the atmosphere is of the order of 100 m.

In the outer region II, the mean velocity, the turbulence energy, and its dissipation, are constant along the height.

This schematization seems wholly admissible for practical simulation. Thus, vertical gradients of the mean velocity in the atmosphere above 100 m are negligible, but represent several percent cent of the gradients in the lower layer. The turbulence intensity is practically constant with height (see /7/). The variation is the dissipation of turbulent energy in the atmospheric layer above 100 m has not yet been reliably measured; the available experimental data are few and contradictory. It is mentioned in /4/ that the dissipation of turbulent energy decreases weakly with height. It is known that in the ground layer of the atmosphere the turbulent energy dissipation decreases linearly with height in the case of neutral stratification.

Naturally, this division into two regions is somewhat arbitrary, and is used only to ensure relative simplicity of the simulation.

We shall now determine the values of the terms $F^{(i)}$ resulting from the introduction of random forces. According to the definition of region II, we assume that the turbulent characteristics and the wind velocity are constant in it, and that the turbulence is isotropic. It follows that all the odd moments are equal to zero, and that the second moments and the flow of energy dissipation are non-diagonal.

From (2) we obtain

$$F_{11}^{(2)^0} = F_{22}^{(2)^0} = F_{33}^{(2)^0} = \frac{2}{3} Q^0 . \tag{19}$$

Similarly, we obtain from (3), (4), and (5):

$$F_{ijk}^{(3)^0} = 0, \tag{20}$$

$$F^{(4)^0} = 4 \frac{Q^{0^2}}{R^0} \tag{21}$$

$$F_i^{(5)^0} = 0.$$

Thus, the scales for $F^{(i)}$ are expressed in terms of the scales for the energy of turbulence and its dissipation.

To obtain the criteria of similitude in our scheme, we must still find what turbulent characteristics can enter into the boundary conditions.

In region II, with constant turbulent characteristics, R, Q, U, these clearly also correspond to the boundary conditions in that part of space.

Region I is uniquely defined by the magnitudes v_* and z_0. We have $v_* \sim R_I^{1/2}$ (R_I is R in region I), and the squares of all the fluctuating velocities are determined by v_* only. Thus, R_I and R_{II} are constant, and the lower part of region II coincides with the upper part of I. Therefore, $R_I = R_{II} = R$.

The isotropy in region II presupposes also the equality of the scales in all directions.

Therefore, according to the theoretical foundations of simulation /11/, of all the dimensionless complexes (11)–(15), the criteria of similitude will be those which include R^0, Q^0, U^0, and the medium parameter ν^0. From (11) we obtain:

$$\frac{R^0}{U^{0^2}} = \text{Ka (the von Karman number)};\qquad (22.1)$$

$$\frac{X^0 U^0}{\nu^0} = \text{Re (the Reynolds number)}.\qquad (22.2)$$

It follows from (12) that

$$\frac{Q^0 X^0}{R^0 U^0} = \text{idem}.\qquad (22.3)$$

From the requirements of geometric similitude, we have

$$\frac{z_0}{X^0} = \text{idem},\qquad \frac{\delta}{X^0} = \text{idem}.\qquad (22.4)$$

The other dimensionless complexes do not give new similitude criteria. If we neglect the influence of molecular viscosity compared to turbulent viscosity, as is usually done, we must exclude Re from the number of criteria of similitude. Then there remain only two criteria (22.1) and (22.3), plus the geometrical requirement (22.4).

We must emphasize most of the dimensionless complexes (11)–(15) are not criteria of similitude, because of the schematization of the impinging atmospheric flow (Figure 1). Thus, in the absence of isotropy in region II, it would have been necessary to introduce different linear scales X_i^0, and the number of criteria of similitude would have increased sharply.

Analysis of the criteria of similitude obtained

It can be shown that the requirement of the proportionality of the turbulent and kinematic linear scales follows directly from the criteria (22.1) and (22.3) obtained.   In fact, by dividing (22.3) by the square root of (22.1), we obtain

$$\frac{Q^0 X^0}{R^{0^{3/2}}} = \frac{X^0}{X^0_{\text{turb}}},\qquad(23)$$

where

$$X^0_{\text{turb}} = \frac{R^{0^{3/2}}}{Q^0}$$

has the meaning of turbulent length scale.

Next, we compare the conclusions drawn from the criteria of similitude (22.1), (22.3), and (22.4), with the experimental data of other authors determined in simulations of the flow past obstructions.

The authors of /22/ compared the velocity field round small obstructions under field conditions and in a wind tunnel. The roughness of the underlying surface in the tunnel was varied for other invariable conditions.   The roughness was determined from the slope of the velocity profile on a semilogarithmic scale.   It was found that for

$$\frac{z_{0M}}{X^0_M} = \frac{z_{0N}}{X^0_N},\qquad(24)$$

where z_{0M}, z_{0N} are the values of the underlying surface roughness in the model and in the field, the velocity field is nearest to its simulation.

This is explained from the point of view of criteria (22.1), (22.3) and (22.4) as follows.   Since the obstructions were small (their height did not exceed several meters, i. e., they were small compared with region I), the characteristics of region I had a strong effect on the impinging flow field, and therefore the roughness had such a strong influence.

Condition (24) alone corresponds to the requirement of geometrical similitude of all the magnitudes with dimensions of length (22.4).

The authors of /26/ and /21/ suggest that the parameter

$$G = \frac{R^{0^{1/2}}}{U^0}.\qquad(25)$$

be taken into account when simulating air flow past ground
obstructions. Another criterion of similitude used in (26) is S:

$$S = \frac{\left(\frac{du}{dz}\right)^0 X^0}{U^0},$$ (26)

where $\left(\frac{du}{dz}\right)^0$ is the characteristic gradient of the mean velocity.

Criterion (26) corresponds to the requirement of conformity to
criterion (22.1), and the geometric similitude of the logarithmic
profiles of the impinging stream velocity in the boundary layer of
the wind tunnel and the ground layer of the atmosphere. In fact,
(26) for a logarithmic profile can be rewritten in the form

$$S = \frac{X^0}{z_0} \frac{v_*^0}{U^0}.$$

The first factor here is equal to one, according to the geometrical similitude, and the second represents the square root of
(22.1), since $v_* \sim R^{1/2}$. Thus, conformity to conditions (25) and (26)
implies only partial conformity to criteria (22.1), (22.3), and (22.4);
criterion (22.3), which includes Q^0, the determining parameter from
region II, is invalid.

Therefore, when criteria (25) and (26) are valid, we should
expect better similitude for low obstructions, where the logarithmic
profile in region I has a decisive influence.

In /21/, as a necessary condition of similitude the authors use not
only criterion (22.1) but also equality of the turbulent Reynolds
numbers, i. e.,

$$\text{Re}_{\text{turb}} = \frac{U^0 X^0}{N^0_{\text{turb}}},$$ (27)

where N^0_{turb} is the characteristic value of the coefficient of eddy
viscosity.

It is easily seen that (27) can be represented as the ratio of the
criteria of similitude (22.1) and (22.3). In fact, if according to (14)

$$N^0_{\text{turb}} = \frac{R^{0^2}}{Q^0},$$

then

$$\text{Re}_{\text{turb}} = \frac{U^0 X^0 Q^0}{R^{0^2}} = \frac{\frac{X^0 Q^0}{R^0 U^0}}{\frac{R^0}{U^{0^2}}},$$ (28)

therefore, (27) is formally a direct corollary of the criteria of similitude (22.1) and (22.3). It is however, important to stress that the author of /21/ does not indicate where N_{turb}^0 is to be measured. Criteria (22.1), (22.3), and (27) cover only the determining parameters of region II, and therefore they cannot ensure similitude in region I, in the lower part of the turbulent stream.

The authors of /19/ list a number of conditions necessary for the similitude of small and intermediate-scale phenomena (these scales are not defined accurately). Most of these conditions coincide with the requirements of geometrical similitude of the logarithmic parts of the boundary layers of the model and nature. Besides, the authors of this paper consider that to simulate the flow past obstructions in the atmosphere it is possible to use even a laminar flow in the wind tunnel. It is claimed (without proof) that the turbulent flow will be similar to the laminar if the turbulent number Re is equal to the molecular number Re.

We believe that this assertion is incorrect, since the equations describing laminar flow and the equations describing turbulent flow are quite different. The theory of similitude specifies similar equations to describe similar processes /11/. Besides, quantitative assessments for the above characteristic values in the atmosphere yield the following values for the eddy and kinematic viscosity:

$$N_{turb}^0 = \frac{R^{0^2}}{Q^0} \sim 10^5 \, cm^2/sec,$$

$$\nu_{kin}^0 \sim 0.1 \, cm^2/sec.$$

It is thus seen that if we ascribe some viscosity N_{turb}^0 to the turbulent atmosphere, it will exceed by several orders the viscosity of laminar flow in the wind tunnel. We can attain similitude for very weak turbulence only.

The only criterion of similitude in /24/ is

$$\frac{Q^0 X^0}{U^{0^3}}. \tag{29}$$

It corresponds again to the combination of criteria (22.1) and (22.3):

$$\frac{Q^0 X^0}{U^{0^3}} = \frac{Q^0 X^0}{R^0 U^0} \frac{R^0}{U^{0^2}}. \tag{30}$$

All the assertions on criterion (27) apply to criterion (29) as well.

In concluding this analysis of the different criteria of similitude, we note that they all reflect only some of the requirements which we believe (see criteria (22.1)–(22.4)) ensure the similitude of the dynamic effect of the obstruction on the impinging turbulent stream. Some of the parameters determining the process are missing, and these criteria also have the drawback that the points of flow at which the different magnitudes should be measured are not defined.

Finally, it can be shown that Taylor's formulas /16/ or /19/ follow from criteria (22.1) and (22.3). According to these formulas the critical Reynolds number for the sphere drag should depend only on the combination

$$\frac{v'}{U^0}\left(\frac{D}{L}\right)^{1/5},$$

(31)

where $v' \sim R^{0^{1/2}}$, D is the sphere diameter, L the characteristic external scale of turbulence.

In fact, if

$$L = \frac{R^{0^{3/2}}}{Q^0},$$

then

$$\frac{R^{0^{1/2}}}{U^0}\left(\frac{X^0Q^0}{R^{0^{3/2}}}\right)^{1/5} = \left(\frac{X^0Q^0}{R^0U^0}\right)^{1/5}\left(\frac{R}{U^2}\right)^{2/5}.$$

(32)

Thus, criteria (22.1, (22.3) are sufficient conditions for the correctness of the Taylor formula, since any combination of criteria of similitude is a criterion of similitude.

It is known that, when simulating body drag in a wind tunnel, the degree of turbulence must be reduced as much as possible. We shall explain this by means of the criteria that we suggested. If we use the expression from (32) for the external scale of turbulence, and take the product of (22.1) and (22.3), which is also the criterion of of similitude $K_{1,2}$, we obtain

$$K_{1,2} = \frac{R^0}{U^{0^2}}\frac{Q^0X^0R^{0^{1/2}}}{R^0U^0R^{0^{1/2}}} = \left(\frac{R^{0^{1/2}}}{U^0}\right)^3\frac{X^0}{L}.$$

(33)

If $X^0 \ll L$, which is practically always the case in the atmosphere for bodies which are not too large (such as an aircraft at great altitudes), then if we consider that $\frac{R^{0^{1/2}}}{U^0} \sim 0.1$, we obtain

$$K_{1,2} \ll 1.$$

(34)

In the wind tunnel the inequality $X^0 \ll L$ is no longer valid, since the value of the external scale of turbulence coincides with the body dimensions. For the condition $K_{1,2} \ll 1$, we must decrease the degree of turbulence.

For criteria (22.1) and (22.3), we should note that they must ensure similitude both inside the boundary layer (with the addition of the natural geometrical parameter of roughness), and in the external stream.

It appears that the external part of the stream must appreciably affect the turbulent conditions of the boundary layer in the wind tunnel. Therefore, to test the correctness of the suggested criteria, we must analyze the results of experiments on the influence of the external conditions on the structure of the boundary layer. This influence of the external flow on 80–90% of the boundary layer thickness is noted in /18/, /20/, /21/. However, the character of this influence has not yet been sufficiently studied. In the large A-6 wind tunnel of the Moscow University Institute of Mechanics Gorlin directed the first experiments on the study of the influence of the degree of turbulence of the external stream on the boundary layer characteristics. The DISA thermoanemometer was used to study the distribution of the degree of turbulence inside the boundary layer for three values of the board roughness and two values of the degree of turbulence of the external stream.

Figure 2 shows the distribution of the degree of turbulence $\varepsilon = \sqrt{\dfrac{v^2}{u^0}}$ that we obtained inside the boundary layer for two extreme values of the plate roughness and two values of the degree of turbulence. Two very important conclusions can be drawn: 1) increase in the degree of turbulence of the external stream leads to a large decrease in the vertical gradient ε (at $\varepsilon_0 = 0.3\%$ the value of ε decreases roughly 20 times when the height changes from 0.2 to 1.0δ, where δ is the boundary layer height, while at $\varepsilon_0 = 4\%$, the value of ε decreases 2.5 times only); 2) increase in ε_0 considerably reduces the influence of the roughness on the value and vertical distribution of the degree of turbulence.

In the same wind tunnel weight experiments were also conducted to study the dependence of the drag of different bodies on the degree of turbulence of the external flow /6/. The results of these experiments indicate that the drag is highly dependent on the degree of turbulence of the external flow.

The experiments listed can be considered as an indirect confirmation of the correctness of our criterion (22.1).

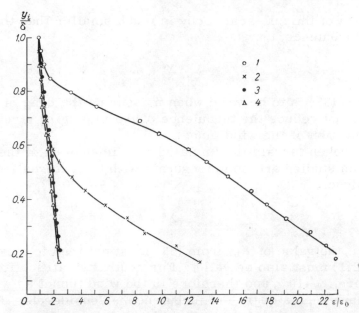

FIGURE 2. Distribution of the degree of turbulence (longitudinal component) in the boundary layer for $U_0 = 10\,\text{m/sec}$:

1) $\epsilon_0 = 0.2\%$ for a rough board; 2) $\epsilon_0 = 0.2\%$ for a smooth board; 3) $\epsilon_0 = 4\%$ for a rough board; 4) $\epsilon_0 = 4\%$ for a smooth board.

Thus, when the proposed method of simulation of the flow past obstructions in the boundary layer of the atmosphere is compared with the earlier methods it is seen that the former is more complete. The earlier criteria can be obtained as particular cases from the criteria of similitude (22.1) and (22.3). But the latter do not follow from the facts and theoretical conclusions given above.

We should note that only two turbulent characteristics R^0 and Q^0 remain as process-determining magnitudes in the suggested simplified model. This simplification of the turbulence description was necessary because of the difficulty of the experimental measurement of the other turbulent characteristics (in both the atmosphere and wind tunnel) in the dimensionless complexes (11)–(15). In a more complete description of the turbulence it is possible to use other criteria from among the dimensionless complexes (11)–(15).

In conclusion, we shall indicate some features of the simulation of obstructions in a wind tunnel.

Different types of experiments are at present being carried out in wind tunnels, mainly on drag measurements. The velocity field and some turbulent characteristics are less frequently measured.

The size of the full-scale body is much smaller than the external scale of turbulence, i. e.,

$$X^0 \ll L. \tag{35}$$

If criterion (33) is to be valid when measuring the drag of the model body, we must reduce the turbulence of the impinging stream in the working section of the wind tunnel.

When broken terrain is simulated the dimensions of the obstruction studied are commensurate with the external scale of the turbulence,

$$X^0 \sim L. \tag{36}$$

Therefore, the value of $K_{1,2}$ from (33) must not tend to zero. Criterion (22.1) must also be valid. For neutral stratification in the atmosphere, $\varepsilon_0 \sim 10\%$, and therefore in the wind tunnel as well we must ensure such a degree of turbulence. Considerable difficulties were overcome at the Moscow State University Institute of Mechanics to create a degree of turbulence in the wind tunnel corresponding to atmospheric turbulence.

The ratio $\dfrac{Q_M}{Q_N}$ between the energy dissipations in the wind tunnel and in the atmosphere varies between 2 and 2,350, according to the results of /19/. High degrees of turbulence correspond to large values of this ratio. We can rewrite criterion (22.3) in the form

$$\frac{Q_N^0 X_N^0}{U_N^{0^3}} = \frac{Q_M^0 X_M^0}{U_M^{0^3}},$$

$$\frac{X_M^0}{X_N^0} = \frac{Q_N^0 U_M^{0^3}}{Q_M^0 U_N^{0^3}}. \tag{37}$$

From (37) it is seen that when $\dfrac{Q_M}{Q_N} \sim 1,000$, a simulation with ratio $\dfrac{X_M^0}{X_N^0} \sim 1,000$ is possible at velocities in the wind tunnel equal to the natural velocities.

If we know the geometry of the obstruction being studied and the height of the boundary layer of the atmosphere, we can obtain geometric similitude between nature and the model, and between the ground layer of the atmosphere and the ground layer in the wind tunnel. But we must first determine the profile of the mean velocity above the board in the absence of a model.

22048

We vary the board roughness until

$$\frac{z_{0N}}{X_N^0} = \frac{z_{0M}}{X_M^0} \quad \text{and} \quad \frac{H_N}{X_N^0} = \frac{H_M}{X_M^0}, \tag{38}$$

where H_N is the height of the ground layer of the atmosphere, H_M the height of the boundary layer in the wind tunnel. Finally, the velocity of the external stream in the wind tunnel (when ϵ_0 is constant in the external streams) must, according to (37), be

$$U_M^0 = U_N^0 \left(\frac{Q_M^0}{Q_N^0} \frac{X_M^0}{X_N^0} \right)^{1/3}. \tag{39}$$

Thus, the estimates given of the magnitudes in the criteria of similitude (22.1), (22.3), and (22.4) indicate the practical possibility of conforming to these criteria when simulating atmospheric streams in existing wind tunnels.

Bibliography

1. Berlyand, M.E. and R.I. Onikul. Fizicheskie osnovy rascheta rasseivaniya v atmosfere promyshlennykh vyb- rosov (Physical Principles for Calculating the Dispersion of Industrial Discharges in the Atmosphere). — Trudy GGO, No. 234, pp. 3—27. 1968.
2. Berlyand, M.E., E.L. Genikhovich, and V.K. Dem'ya- novich. Nekotorye aktual'nye voprosy issledovaniya atmosfernoi diffuzii (Some Topical Problems of the Investigation of Atmospheric Diffusion). — Trudy GGO, No. 172, pp. 3—22. 1965.
3. Berlyand, M.E., E.L. Genikhovich, and O.I. Kurenbin. Vliyanie rel'efa na rasprostranenie primesi ot istochnikov (Influence of the Relief on the Propagation of a Contami- nant from a Source). — Trudy GGO, No. 234, pp. 28—44. 1968.
4. Byzova, N.L., V.N. Ivanov, and S.A. Morozov. Turbu- lentnye kharakteristiki skorosti vetra v pogranichnom sloe atmosfery (Turbulent Characteristics of the Wind Velocity in the Boundary Layer of the Atmosphere). — In: Atmosfernaya turbulentnost' i rasprostranenie radiovoln. Moskva, "Nauka." 1967.

5. Gorlin, S. M. and I. M. Zrazhevskii.   Izuchenie obtekaniya modelei rel'efa i gorodskoi zastroiki v aerodinamicheskoi trube (Study of Flow past Models of Relief and Town Buildings in a Wind Tunnel).— Trudy GGO, No. 234, pp. 45—59. 1968.

6. Gorlin, S. M. and G. E. Khudyakov.   Vliyanie nachal'noi turbulentnosti potoka na aerodinamicheskie kharakteristiki ploskoi plastinki (Influence of the Initial Flow Turbulence on the Aerodynamic Characteristics of a Flat Plate).— Nauchnye Trudy Instituta Mekhaniki MGU, No. 4. 1970.

7. Gurevich, A. S. et al.   Empiricheskie dannye o melkomasshtabnoi strukture atmosfernoi turbulentnosti (Empirical Data on the Small-Scale Structure of Atmospheric Turbulence).— In: Atmosfernaya turbulentnost' i rasprostranenie radiovoln. Moskva, "Nauka." 1967.

8. Davydov, B. I.   K statisticheskoi dinamike neszhimaemoi turbulentnoi zhidkosti (Statistical Dynamics of an Incompressible Turbulent Liquid).— DAN SSSR, Vol. 127, No. 4. 1959.

9. Davydov, B. I.   K statisticheskoi teorii turbulentnosti (Statistical Theory of Turbulence).— DAN SSSR, Vol. 127, No. 5. 1959.

10. Davydov, B. I.   K statisticheskoi dinamike neszhimaemoi turbulentnoi zhidkosti (Statistical Dynamics of an Incompressible Turbulent Liquid).— DAN SSSR, Vol. 136, No. 1. 1961.

11. Kirpichev, M. V.   Teoriya podobiya (Theory of Similitude). Moskva, Izdatel'stvo AN SSSR. 1953.

12. Gol'tsberg, I. A. (Editor).   Mikroklimat kholmistogo rel'efa i ego vliyanie na sel'skokhozyaistvennye kul'tury (Microclimate of Hilly Terrain and its Influence on Agricultural Crops). Leningrad, Gidrometeoizdat. 1962.

13. Monin, A. S.   O svoistvakh simmetrii turbulentnosti v prizemnom sloe vozdukha (The Properties of the Symmetry of Turbulence in the Ground Air Layer).— Izvestiya AN SSSR, Fizika Atmosfery i Okeana, Vol. 1. 1965.

14. Monin, A. S. and A. M. Yaglom.   Statisticheskaya gidromekhanika.   Mekhanika turbulentnosti. Ch. 1 (Statistical Hydromechanics. Mechanics of Turbulence. Part 1). Moskva, "Nauka." 1965.

15. Monin, A. S. and A. M. Yaglom.   Statisticheskaya gidromekhanika. Mekhanika turbulentnosti. Ch. 2 (Statistical Hydromechanics. Mechanics of Turbulence. Part 2). Moskva, "Nauka." 1967.

16. Rotta, I. K.   Turbulentnyi pogranichnyi sloi v neszhimaemoi
    zhidkosti (The Turbulent Boundary Layer in an Incompres-
    sible Liquid). Leningrad, "Sudostroenie." 1967.
17. Sapozhnikova, S. A.   Mikroklimat i mestnyi klimat (Micro-
    climate and Local Climate). Leningrad, Gidrometeoizdat.
    1950.
18. Khintse, I. O.   Turbulentnost' i ee mekhanizm i teoriya
    (Turbulence and its Mechanism and Theory). Moskva,
    Fizmatgiz. 1963.
19. Cermak, J. E. et al.   Simulation of Atmospheric Motion by
    Wind-Tunnel Flows. — Technical Report. Fluid Dynamics
    and Diffusion Laboratory; College of Engineering, Colo-
    rado State University. May. 1966.
20. Grant, H. L.   The Large Eddies of Turbulent Motion. — J. Fluid
    Mech., Vol. 4, pp. 149—190. 1958.
21. Inoue, E.   The Structure of Surface Wind. — Bull. Natn. Inst.
    Agric. Sci., Tokyo, Series A, No. 2, p. 93. 1952.
22. Jensen, M. and N. Franck.   Model-Scale Tests in Turbulent
    Wind. Part 1. Copenhagen. 1963.
23. Millikan, C. B. and A. L. Klein.   The Effect of Turbulence. —
    Aircr. Engng., p. 169. August. 1933.
24. Nemoto, S.   Similarity between Natural Local Wind in the
    Atmosphere and Model Wind in a Wind Tunnel. — Pap. Met.
    Geophys., Tokyo, Vol. 19, No. 2, pp. 131—230. 1968.
25. Taylor, G. I.   Statistical Theory of Turbulence. Part 5.
    Effect of Turbulence on Boundary Layer. Theoretical
    Discussion of Relationship between Scale of Turbulence
    and Critical Resistance of Spheres. — Proc. R. Soc.,
    Lond., A 151, pp. 307—317. 1936.
26. Scorer, R. S.   The Limitations of Wind Tunnel Experiments
    in the Study of Airflow around Buildings. — Int. J. Air Wat.
    Pollut., Vol. 7, pp. 927—931. 1963.
27. Wyld, H. W.   Formulations of the Theory of Turbulence in an
    Incompressible Fluid. — Ann. Phys., Vol. 14, No. 32. 1961.

EXPERIMENTAL STUDIES OF THE DISPERSION IN THE ATMOSPHERE OF COLD VENTILATION DISCHARGES OF RAYON PLANTS

B. I. Vdovin, I. M. Zrazhevskii,
T. A. Kuz'mina, R. I. Onikul,
A. A. Pavlenko, G. A. Panfilova,
G. P. Rastorgueva, B. V. Rikhter

The laws of propagation in the atmosphere of discharges from plants with cold and low discharges were until recently little studied. The standards in force in the Soviet Union for calculating the dispersion of contaminants impose many restrictions on their application to chemical and other industries.

In 1966—1968, M. E. Berlyand organized four expeditions of the Voeikov Main Geophysical Observatory to study the dispersion in the atmosphere of hydrogen sulfide and carbon disulfide from the rayon plants in Cherkassy, Krasnoyarsk, and Balakovo (two expeditions). The following participated in the studies: the State Institute for the Planning of Synthetic Fiber Establishments, the Erisman Moscow Institute of Hygiene, the Odessa Hydrometeorological Institute, and local branches of the Administration of Hydrometeorological Services. The staffs of the plants being studied, and in particular, their central laboratories and ventilation subsections, were of great help.

The main aim of the expeditions was to obtain experimental results to develop and corroborate the method for calculating the dispersion in the atmosphere of cold discharges. The CS_2 and H_2S components were studied.

In and around the plants there were sources (such as thermal power plants) of discharge of sulfur dioxide in the atmosphere. The concentrations of this gas in the atmosphere were not measured, and it was removed before sampling.

Most of the discharged hydrogen sulfide and carbon disulfide was directed to 2—3 ventilation stacks of height 120 m. Considerable amounts of these substances were also discharged in the atmosphere through air shafts, fans, baffles of the industrial premises, and through the sewage in special settling tanks located at some distance from the enterprises.

Since the temperature in the plants varied over the narrow range of 20—40°, in summer in daytime the discharges differed little in temperature from the surrounding air, and could therefore be called cold discharges in accordance with standard terminology.

In Cherkassy the expeditions were conducted between 20 July and 5 September, 1966, in Krasnoyarsk between 27 February and 31 March, 1967, in Balakovo between 23 July and 3 September, 1967, and between 13 August and 9 September, 1968.

In Cherkassy the plant was built on the outskirts of a town with mainly one- and two-storied buildings.   The area for meteorological and aerological observations was under roughly the same conditions. The region within a radius of 10—15 km from the plant was relatively flat, and its main feature was the large Kremenchug reservoir on the Dniepr, with an area of more than 2,000 km^2, adjoining the town.

In Krasnoyarsk the source was located in the midst of residential areas with two- and three-storied buildings predominating, in the valley of the Yenisei River which was very wide within the town limits.   The sampling region, within a radius of 5—8 km from the plant, had a relatively flat underlying surface.   Meteorological observations near the plant were relatively limited.   The main meteorological and aerological observations were conducted at the Yemelyanovo meteorological station, located 30 km from Krasno- yarsk in the Yenisei River valley, in a flat area with low hills. When describing the underlying surface, we must also note that due to the anomalously early arrival of spring, the snow cover melted in the period between the 10th and 17th of March.

The Balakova rayon plant was located at some distance from the town of Balakovo.

The area for the meteorological and aerological observations was located on an open, slightly hilly, plain.

Method of sampling and sample analysis

The distribution of contaminants in the ground air layer was characterized by their single concentrations.

On all the expeditions, sampling was carried out during daytime at a height of about 1.5 m, at first under the plant jet.   The daily period of sampling was 5—7 hours.   Samples were taken hourly for 20—30 minutes.   To study a wider range of meteorological con- ditions, including ground and high temperature inversions, a sliding sampling curve was used.   According to this curve, on all the ex- peditions except the Krasnoyarsk expedition, sampling was conducted between 14 and 21 hours on the first and fourth days of the week, between 9 and 16 hours on the second and fifth days, between 6 and 13 hours on the third and sixth days. With this working curve it was possible to allow more completely for the influence of the daily variation in the meteorological factors on dispersion of contaminants. On the Krasnoyarsk expedition sampling was conducted six times weekly from 9 to 16 hours.

The gases discharged from high stacks were invisible on summer expeditions. In Krasnoyarsk, at low air temperatures the jet from the stacks extended to distances of several hundred meters as a result of the condensation of the water vapor contained in the discharged gases.

The discharges from the hundreds of baffles, shafts, and other small sources distributed over the territory were difficult to measure, and varied rapidly with time, depending on many factors not taken into account in practice. The characteristic of air pollution near these sources was not a direct problem.

Accordingly, sampling was conducted at 0.5—1 km from the center of high ventilation stacks, on their lee side. The basic observations were conducted starting at a distance of 2 km, where, as shown by the calculations in /2/, a relatively homogeneous common gas jet was formed, with maximum concentrations on the axis in the direction of the wind and passing through the plant center. The axis of the invisible jet was determined tentatively from the characteristic smell with allowance for wind direction and the smoke jet of the TPP.

The sampling points were located at standard distances from the plant center (0.5, 1, 2, 3, 4, 5, 6, 7, 8, 9, 10, 12, and 15 km). The position of the sampling point was determined from maps of the region; if no clear reference points were available in the near vicinity, it was determined from the readings of range finders and car speedometers.

In the Cherkassy and Balakovo expeditions of 1967, 15 km wide zones were studied, in the Balakovo expedition of 1968, a 12 km wide zone, and in the Krasnoyarsk expedition a 8 km wide zone.

At each distance on the jet axis some of the samples were taken by means of LK-1 electric suction pumps, capable of taking two hydrogen sulfide and two carbon disulfide samples simultaneously. However, electric suction pumps required heavy power supply units; accordingly, in Cherkassy and Balakovo in 1967, less perfect water suction tanks were used for sampling.

The sampling in Cherkassy and Balakovo was performed simultaneously at 3—4 ranges at several points on each. In Krasnoyarsk,

samples were taken simultaneously by means of electric suction
pumps at two points only at different distances, due to the more
difficult conditions of the cold season.

An automatic machine with electric suction pump was usually
placed on the jet axis, and stations with water tanks at its sides.
The distance between adjacent stations on one range was not
accurately fixed, and varied between 50 and 400 m, depending on the
sampling conditions and other factors.   In particular, at near
ranges this distance was usually less than at far range, where the
concentration field is more homogeneous in the direction transverse to the jet.

In the first Balakovo expedition, the concentrations of hydrogen
sulfide on the jet periphery generally lay within the sensitivity
limits of the measurement method, and therefore in the second expedition they were measured only along the assumed jet axis from
one sample per range at any given moment.   The results are given
in Table 1.

TABLE 1. Number of samples of hydrogen sulfide and carbon disulfide taken in expeditions
at different distances from the center of the industrial area

Distance from center of industrial area (km)	Expeditions							
	Cherkassy		Krasnoyarsk		Balakovo			
					1967		1968	
	CS_2	H_2S	CS_2	H_2S	CS_2	H_2S	CS_2	H_2S
0.5	62	38	34	21	91	41	171	36
1	320	137	174	91	268	129	257	48
2	680	268	148	79	527	201	193	54
3	252	118	116	60	311	124	229	49
4	299	114	54	27	326	123	321	69
5	250	103	54	27	164	46	212	44
6	143	58	48	25	191	61	182	44
7–8	55	25	46	23	389	110	494	100
9–15	302	126	–	–	295	66	445	91
Total	2,363	987	674	353	2,562	904	2,504	535

A total of more than 8,000 carbon disulfide samples were found
suitable for processing, and about 3,000 hydrogen sulfide samples.

In Krasnoyarsk, samples were also taken by means of aircraft
and helicopters directly in the jets of the ventilation stacks.   The
results are given in /7/.

In Balakovo, in 1967 a stationary observation station located 3 km to the northwest of the plant was set up, and it took 200 samples each of hydrogen sulfide and carbon disulfide. In addition, in the course of 5 days air was sampled at 100, 250, and 500 m from the lake (on the lee side), . The lake lies 5 km from the plant, and sewage wastes are discharged into it. A total of 250 carbon disulfide samples and 50 hydrogen sulfide samples were taken.

Sampling was performed in U-shaped absorption devices. The samples were analyzed on the day they taken or on the following day. A standard method for determining carbon disulfide in the air was used, and the value 0.03 mg/m^3 was taken as the maximum admissible concentration (MAC). This method is based on absorption of carbon disulfide by an alcoholic solution of diethylamine, and the determination of the concentrations from the intensity of the yellow-brown color of copper diethylcarbamate /10/. The sensitivity of the method is 0.03 mg/m^3. The method is not specific in the presence of hydrogen sulfide. Accordingly, on the first expedition a device for absorbing hydrogen sulfide was placed in front of the absorbing device with diethylamine solution. Subsequently, a more convenient and reliable sampling method was developed for determining the concentrations of carbon disulfide in the air /11/. Special cartridges with a solid chemosorbent have been suggested for removing hydrogen sulfide and sulfur dioxide.

The hydrogen sulfide (MAC equal to 0.008 mg/m^3) was determined on the first expedition by a standard method /10/, based on the absorption of hydrogen sulfide by an alkaline solution of sodium arsenite.

The sulfo salt formed interacts with an acid solution of silver nitrate to yield a yellow-brown colloidal solution of silver sulfide. The method is specific, but its sensitivity is low, 0.012 mg/m^3. Later, a more sensitive method was introduced, based on the absorption of hydrogen sulfide by a suspension of cadmium hydroxide, and determination by means of the methylene blue formed with paraaminodimethylaniline /1/. The sensitivity is 0.006 mg/m^3. The sulfur dioxide in the sampled air was removed by means of special cartridges with a suitable chemosorbent.

At high air temperatures the absorbents were enclosed in special baths with cold water to prevent excessive evaporation of the solution.

Meteorological and aerological observations

On all the expeditions, round-the-clock detailed meteorological and aerological observations were carried out. The set of observations was increased appreciably during sampling periods. The program of meteorological observations in the ground layer included: measurement of the temperature and humidity of the air and soil, the speed and direction of the wind, the precipitations and pressure, visual observations of cloudiness and specific weather phenomena.

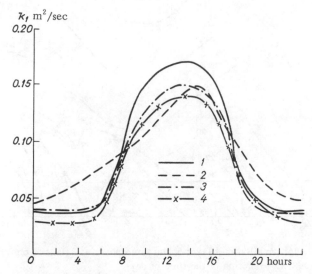

FIGURE 1. Mean curve of the coefficient of eddy diffusion as a function of the time of day at a height of 1 m, calculated from gradient observations in the course of different expeditions:

1) Cherkassy; 2) Krasnoyarsk; 3) Balakovo, 1967; 4) Balakovo, 1968.

Gradient observations of the wind velocity, the temperature and humidity of the air in the layer up to 17 m (8 m in Krasnoyarsk) were carried out on telescope towers. From these observations the coefficients of eddy diffusion, and the turbulent flows of contaminants, heat, and humidity, were calculated. As an example, in Figure 1, mean values of the coefficient of eddy diffusion are given as a function of the time of day for all the expeditions.

In summer expeditions, the soil humidity and its temperature were measured at the surface, and to a depth of 20 cm. Evaporation from the soil surface was measured by an evaporimeter. For closure of the heat balance of the underlying surface, the radiation balance components (direct, scattered and reflected solar radiation, and effective radiation of the underlying surface) were measured by an actinometer. Independent additional information on the coefficient of vertical eddy diffusion was obtained from the heat balance observations.

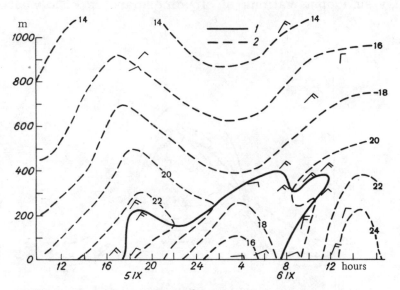

FIGURE 2. Bottom kilometer layer of the atmosphere at Balakovo on 5—6 September 1968, as a function of the time of day:

1) inversion boundary; 2) isotherms.

The wind direction was recorded on all the expeditions by M-12 and M-45 anemobiagraphs. The results were used to study dispersion in the direction of the wind, which is an important characteristic of atmospheric diffusion in a direction transverse to the jet.

The regular aerological observations included measurements of the air temperature and humidity in the boundary layer of the atmosphere up to heights of 1—1.5 km by means of YaK-12 light aircraft, MI-1 helicopters, and radiosondes. Radiosondes of types A-58, A-22—IV, A-22-VII launched at a low velocity were used for radio sounding /5, 6, 8/.

The wind direction and speed at different altitudes in the boundary layer of the atmosphere were determined by means of observations of the pilot balloons and radiosondes from one station and from the base. From the observations carried out from the base it was possible to obtain some parameters characterizing turbulence in the boundary layer /9/.

The program of aircraft sounding included recording of overloads and temperature fluctuations, in order to calculate the vertical component of the coefficient of eddy diffusion. In Krasnoyarsk, the characteristics of eddy diffusion were studied by means of aircraft and helicopter, which investigated the visible contours of the jets emitted from high ventilation stacks and the distribution of concentrations in them.

More detailed information on the carrying out of the total aerological observations is given in /5/ and /6/.

From the aerological data of all the expeditions, time sections of the boundary layer of the atmosphere, visually representing the evolution of the inversion, the layers of higher instability, the wind regime, etc., were plotted and a more exact determination of the aerological structure in the period between observations was possible (Figure 2).

The isotherms and the inversion boundaries are drawn in the figure. On the corresponding altitudes are also shown the wind direction and speed. One long "feather" corresponds to a wind velocity of 2 m/sec, and a short one to 1 m/sec

Characteristic of the weather conditions

Different numbers of samples were taken at different moments of time; however, calculations showed that the specific number of samples of the components taken under different meteorological conditions was roughly the same, and accordingly we give their mean values in Table 2 and in Figures 3 and 4. These figures are self-explanatory. We should note that the largest number of samples corresponds to convective weather conditions with air temperatures of 15–30° and wind velocities of 1–6 m/sec.

TABLE 2. Number of samples (in %) for different types of temperature stratification in the 0–500 m layer

Locality	Stratification			
	unstable	equilibrium	high inversion	ground inversion
Cherkassy	67	–	21	12
Krasnoyarsk	30	40	19	11
Balakovo				
1967	51	11	22	16
1968	54	8	20	18

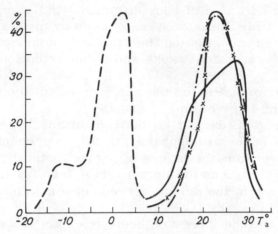

FIGURE 3. Number of samples (average percent for carbon disulfide and hydrogen sulfide) taken at different air temperatures T_a^0 at a height of 2 m. Legend as in Figure 1.

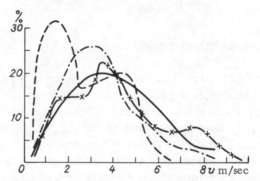

FIGURE 4. Number of samples (average percent for carbon disulfide and hydrogen sulfide) taken at different wind velocities u at the height of the wind vane. Legend as in Figure 1.

Due to the anomalously early start of spring in Krasnoyarsk, relatively complete data were obtained on that expedition for temperatures above -10°, -15°, and mainly between -5 and +5°. The wind was somewhat weaker than in summer.

By using a sliding sampling curve, accompanied by more frequent meteorological and aerological observations, it was possible to obtain extensive material on the pollution of the atmosphere for all basic types of stratification of the atmosphere: unstable stratification, equilibrium conditions, and high and ground temperature inversions. As expected, most of the summer expedition samples corresponded to unstable stratification, although under winter-spring

conditions also this type of stratification was frequently observed
in Krasnoyarsk. Quasi-equilibrium conditions play a significant
part in Krasnoyarsk only. The frequency of ground and high temperature inversions on all the expeditions did not differ greatly.

Evaluation of the discharge parameters

A study of the discharge parameters necessary for calculating
the distribution of concentrations in the atmosphere is of great
importance. Such studies have been carried out in different
directions, including a study of the available literature, generalization of the results of work of similar industries, and direct
studies on plants during the expedition periods. Giproiv's
workers, Menis and Shmerling, carried out these investigations.
The pollution of the atmosphere by the rayon industry is due to
losses of carbon disulfide, used in one of the primary stages of
production for processing alkali cellulose. In this process cellulose xanthate is formed, and its alkaline solution is called "viscose."
The amount of carbon disulfide used to xanthate cellulose can be
exactly calculated. The total discharge of carbon disulfide into the
atmosphere cannot be higher than this value. This fact was used
in a critical assessment of the value of the total release of carbon
disulfide, based on direct measurements of the concentration in
stacks, fans, etc., and it became possible to establish a scientifically
based specific balance of carbon disulfide and hydrogen sulfide.
We shall briefly describe the discharges of these components at different stages of the industrial process in the enterprises studied /12,13/.
Even during the xanthation of cellulose, several percent of the
total amount of carbon disulfide was lost to the atmosphere; this
loss occurred through the low ventilation shafts connected to the
local ventilation of the chemical shops.
In the next stage in the manufacturing process, the viscose was
kept for a long time in tanks, and a large amount of thiocarbonates
was formed. During the spinning of the viscose, hydrogen sulfide
was evolved from the thiocarbonates. The largest amounts of
carbon disulfide and hydrogen sulfide are formed during the spinning
and in part in the finishing shops, under the covers of the machines.
Considerable amounts of these components are also released at the
acid stations during regeneration of the solutions of the precipitation and finishing baths.
Great attention has been paid in the enterprises to sealing the
machines and mounting local ventilation pumps connected to the gas

purification or to high stacks at all parts of the manufacturing process in which considerable amounts of carbon disulfide and hydrogen sulfide are released.

The centralization of the most polluted air in high stacks is efficient for two reasons. Firstly, the natural dispersive properties of the atmosphere are used to the full. Secondly, optimum pre-requisites are created for the efficient operation of the complex and expensive gas purification equipment, which in turn requires high concentrations of carbon disulfide and hydrogen sulfide.

The little polluted ventilation discharges were also connected as far as possible to the high stacks after they had been scrubbed. This scarcely changed the value of the total discharge, while the ascent of the jet increased, and the concentration in the ground layer decreased as a result of the increase in the velocity of discharge in the atmosphere. It is particularly effective, from this point of view, to connect to high stacks, after gas scrubbing, the little polluted but very large volumes of discharge of intensified ventilation switched on in the working places when the spinning and finishing machines are uncovered.

In spite of all the measures taken in the plants during the spinning and finishing of the fiber, the regeneration of the precipi-tation baths and other operations, a certain amount of carbon disul-fide and hydrogen sulfide was discharged into the atmosphere throug through low sources located near the building roofs.

At different stages of the manufacturing process, considerable quantities of carbon disulfide and hydrogen sulfide were discharged with the industrial runoff in the sewage, through which they reached sedimentation basins located at some distance from the enterprise. Some of these substances formed nonvolatile compounds in the sedimentation basins, and the rest penetrated into the atmosphere.

Menis and Shmerling studied and generalized the available material on losses of carbon disulfide and hydrogen sulfide at different stages of the manufacture of rayon. It was found that the existing results have a very high scatter, and are at times even contradictory. Nevertheless, they were able to establish a fairly typical unit balance.

It follows from the balance that about 60% of the carbon disulfide used is released in such a way that it can be directed to gas scrub-bing or to high stacks. About 10% of the carbon disulfide is carried by industrial runoffs in the sewage. The amount of low discharges of carbon disulfide in the atmosphere is very dependent on the hermetic sealing of the spinning and finishing machines, the connection of the local pumps of acid stations to the high stacks, etc. On an average, the low discharges of carbon disulfide can be considered to be from 12—15% of the total amount of carbon disulfide consumed to xanthate the viscose. About 15% of the carbon disulfide is transformed into hydrogen sulfide during the process of ageing of the viscose, with two molecules of hydrogen sulfide formed from each molecule of carbon disulfide. If we consider the molecular weights of carbon disulfide and hydrogen sulfide (76 and 34), we find that about 0.9 g of hydrogen sulfide is formed from 1 g of carbon disulfide. The ratio between low and high discharges of hydrogen sulfide depends only on the measures taken to hermetically seal the machines and on other industrial factors. The balance established shows that when the most complete measures are taken to reduce the low discharges of hydrogen sulfide, these discharges form about 4—5% of the total amount of hydrogen sulfide produced. If the local ventilation of the acid stations is not connected to the high stacks, the contribution of the low discharges of hydrogen sulfide may reach 18—20%. The amount of hydrogen sulfide in the discharges in the sewage is roughly the same as for the carbon disulfide.

On all the expeditions the amount of air discharged from the high stacks and the concentrations of carbon disulfide and hydrogen sulfide in it were measured. The total discharges of these components were calculated from the measurements. However, while the volumes of air varied little during one expedition and were determined with sufficient accuracy from the output of the operating ventilation installations, the measurement of the concentrations of discharged gases in a stack connected to a large number of gas outlets with sharply differing pollution levels proved to be a very complex problem.

In Cherkassy the concentrations of carbon disulfide and hydrogen sulfide were determined for the first time, and mainly for methodological purposes. The results of these studies contradicted the established unit balance. In later expeditions it was possible to achieve a satisfactory agreement between the discharges in high stacks according to the results of direct measurements and the unit balance of carbon disulfide and hydrogen sulfide.

Gas scrubbing was carried out in Balakovo only. Carbon disulfide scrubbing was used in 1967 in one of the plants there, but not daily and not very efficiently. In 1968, there was no gas scrubbing. Hydrogen sulfide scrubbing was efficient during both expeditions, with a mean efficiency of 80—85%; simultaneously, the gas entrained about 15—20% of carbon disulfide. The mean efficiency of the gas scrubbing was determined by measuring the concentration of harmful substances in the gas discharges before and after the gas passed the scrubbing installation. This characteristic of the efficiency of

the gas scrubbing installation is more convenient than measurement
of the residual concentration of harmful components after the gas
has passed the scrubbing installation.

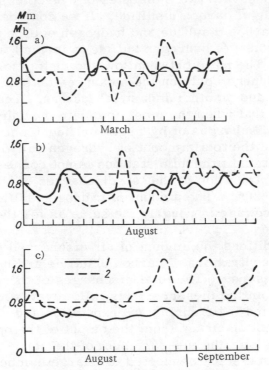

FIGURE 5. Comparison of the total discharges of
carbon disulfide (1) and hydrogen sulfide (2) in a
stack, calculated from the unit balance and the
concentrations of these substances in the gas
passages:

a) Krasnoyarsk, 1967; b) Balakovo, 1967; c) Balakovo 1968.

The ratio M_m/M_b for both components for Krasnoyarsk and
Balakovo is plotted in Figure 5. Here M_m is the measured amount
of harmful substances discharged into the atmosphere on a given
day, M_b the corresponding value calculated from the unit balance
of carbon disulfide and hydrogen sulfide, allowing for the mean
efficiency of the gas scrubbing. The total amount of discharges for
all the plants was used. The agreement between the balance calculations and the measurements must be considered satisfactory.

The height of the stacks in the plants studied was 120 m, and
their diameter 7 m.

The velocity of exit of polluted air w_0 into the atmosphere was determined from the volume of discharges V_1 and the diameter D of the opening of the stack from the formula $w_0 = \frac{4V_1}{\pi D^2}$. In Cherkassy the value of w_0 was about 12 m/sec, in Krasnoyarsk 8 m/sec, and in Balakovo 6 m/sec.

In one of the Balakovo plants, for technical reasons the gases were discharged at a height of 50 m through an opening 10 m in diameter with a velocity of 3 m/sec.

At their exit into the atmosphere the discharges were at a temperature near to that of the industrial premises (20–40°). Therefore, during the entire cold season and a considerable part of the warm, especially at night, the excess temperature of the discharges ΔT was 40–50° or more. In summer, the excess temperature ΔT was small, and in daytime sometimes approached zero.

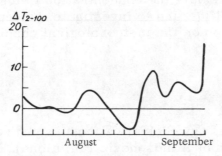

FIGURE 6. Mean excess temperature of the discharges from a stack with reference to air at the level of the stack opening

Figure 6 shows the mean difference between the temperature of the discharges of one of the plants and the air temperature at the level of the stack opening during the sampling period, i.e., during daylight, for Balakovo, 1968.

Each plant had a large number of low discharge sources distributed over the roofs of the industrial premises, ventilation shafts, different types of fans, etc. Some discharges escaped through windows and doors also. The low discharges represented a considerable part of the total discharges.

The total discharges from low sources were studied in 1968 in Balakovo only, where the amount of discharges from 288 sources, and the velocity of exit of the exhaust gases and their temperature were measured three times during the expedition. Considerable discharges were recorded from the exhaust installations of the spinning plants, the acid stations, etc. The total amount of low discharges of carbon disulfide and hydrogen sulfide agrees satisfactorily with the discharges calculated according to the unit balance. Substantial differences between the discharges from similar industries, the small amount of carbon disulfide discharged from the cellulose xanthating shops, point to the necessity of improving the methodology of such studies and of continuing them.

The range of variation in the parameters of the low sources in Balakovo was: height of discharges $H_i = 8-20$ m, opening diameter $D_i = 0.2-2.2$ m, discharge temperature $T_{gi} = 15-40°$, velocity of exit of the ventilation gases $\omega_{0i} = 3-20$ m/sec. The average values of these parameters, weighted according to the contribution of the sources to the pollution of the air basin, are equal to: $H_i = 15$ m, $D_i = 1$ m, $T_{gi} = 25°$ and $\omega_{0i} = 5$ m/sec.

In conclusion we note that, in spite of the considerable difficulties associated with the organization of studies of the rayon plants, the results enable us to calculate the concentration fields of hydrogen sulfide and carbon disulfide, and to investigate the dependence of the atmospheric pollution on the meteorological conditions /2, 4/.

Bibliography

1. Alekseeva, M.V. Opredelenie atmosfernykh zagryaznenii (Determination of Atmospheric Pollutions). Moskva, Medgiz. 1963.

2. Baikov, B.K. et al. Eksperimental'naya proverka metodik rascheta rasseivaniya v atmosfere kholodnykh vybrosov na materialakh obsledovaniya predpriyatii iskusstvennogo volokna (Test of the Method for Computing the Dispersion in the Atmosphere of Cold Discharges on Materials from the Study of Rayon Plants). See this Collection.

3. Berlyand, M.E. and R.I.Onikul. Fizic..eskie osnovy rascheta rasseivaniya v atmosfere promyshlennykh vybrosov (Physical Principles of the Calculation of the Dispersion of Industrial Exhausts in the Atmosphere). — Trudy GGO, No. 234, pp. 3—27. 1968.

4. Berlyand, M.E. and R.I.Onikul. K obobshcheniyu teorii rasseivaniya promyshlennykh vybrosov v atmosferu (Generalized Theory of Dispersion of Industrial Exhausts in the Atmosphere). See this Collection.

5. Vdovin, B.I. and V.G.Voloshin. O chastote i tochnosti temperaturnogo zondirovaniya pogranichnogo sloya

atmosfery (The Accuracy and Frequency of Temperature
Sounding of the Boundary Layer of the Atmosphere). —
Trudy GGO, No. 238. 1969.

6. Voloshin, V. G.   K voprosu o primenenii radiozondov dlya
issledovaniya pogranichnogo sloya atmosfery (Utilization
of Radiosondes in the Study of the Boundary Layer of the
Atmosphere). — Trudy GGO, No. 234, pp. 223—230. 1968.

7. Eliseev, V. S.   Issledovanie struktury dymovoi strui i opredelenie koeffitsienta turbulentnogo peremeshivaniya po
vertikal'nomu raspredeleniyu kontsentratsii (Investigation
of the Smoke Jet Structure and Determination of the
Coefficient of Turbulent Mixing from the Vertical Distribution of Concentrations). — Trudy GGO, No. 234, pp. 95—99.
1968.

8. Efimov, P. A.   Opyt zondirovaniya nizhnego trekhkilometro-
vogo sloya atmosfery radiozondom A-58 (Results of
Sounding the Bottom 3 km Layer of the Atmosphere by the
A-58 Radiosonde). — Trudy TsAO, No. 74. 1966.

9. Zaitsev, A. S.   Issledovanie kharakteristik turbulentnosti s
pomoshch'yu sharov-pilotov (Investigation of the Turbulence Characteristics by Means of Pilot Balloons). —
Trudy GGO, No. 238. 1969.

10. Instruktivno-metodicheskie ukazaniya po organizatsii issledo-
vaniya zagryazneniya atmosfernogo vozdukha (Instructions
Regarding the Organization of Studies on the Pollution of
Atmospheric Air). Moskva, Medgiz. 1963.

11. Pavlenko, A. A. and T. A. Kuz'mina.   K metodam opredeleniya serovodoroda i serougleroda (Methods of Determination of Hydrogen Sulfide and Carbon Disulfide). — Trudy
GGO, No. 234, pp. 188—195. 1968.

12. Rogovin, Z. A.   Osnovy khimii i tekhnologii proizvodstva
khimicheskikh volokon [V 2-kh t] (Principles of the
Chemistry and Technology of the Manufacture of Artificial
Fibers [In Two Volumes]). 3rd edition. Moskva-Lenin-
grad, "Khimiya." 1964.

13. Ryauzov, A. N.   Tekhnologiya khimicheskikh volokon
(Technology of Artificial Fibers). Moskva, "Vysshaya
shkola." 1964.

TEST OF THE METHOD FOR COMPUTING
THE DISPERSION IN THE ATMOSPHERE
OF COLD DISCHARGES ON MATERIALS
FROM THE STUDY OF RAYON PLANTS

B. K. Baikov, R. S. Gil'denskiol'd,
I. M. Zrazhevskii, R. I. Onikul,
G. A. Panfilova

Theoretical studies /1/ aimed at expanding the applicability of
the standard method for calculating the dispersion of industrial
discharges in the atmosphere in use in the USSR /5/ have recently
been carried out at the Voeikov Main Geophysical Observatory.
A procedure was evolved /1/, including /5/ as a particular case,
and extended also to a wide class of sources with cold and weakly
heated discharges, and also with low dangerous wind velocities.
It became possible to conduct scientifically based detailed calculations under unfavorable meteorological conditions for exhaust
discharges from industrial premises and engineering installations,
both when they are centralized in high stacks and when they are
discharged from a large number of low sources.

It was found that in the latter case tentative calculations can be
performed on the basis of the total discharges; this is especially
important in view of the absence of sufficiently complete data on
the discharge parameters of the different low sources of the plants
being designed.

The calculations are simplified by replacing the parameters of
the low sources by their mean values ascribed to one source
located at the center of the industrial area. It was found possible
to conduct the calculations on the basis of the total discharges for
regions located at 1–2 km or more from the industrial area.

In 1966–1968, the Main Geophysical Observatory, together with
the Moscow Erisman Scientific-Research Institute of Hygiene, the
State Institute for the Design of Rayon Plants, and some other
organizations, organized four expeditions in the area of plants producing viscose rayon, and discharging into the atmosphere carbon
disulfide (CS_2) and hydrogen sulfide (H_2S): in Cherkassy (1966),
Krasnoyarsk (1967), and Balakovo (1967 and 1968). The expeditions
are described in /2/.

In these plants the contaminants were discharged through 120 m high ventilation stacks and through numerous low ventilation shafts, exhausts, baffles, and also to a certain extent from the sedimentation basins of the industrial sewerage.

The procedure used to process the results on the concentrations of the contaminants under the plant jet was similar to that used earlier to analyze the observations in the area of the thermal power plants /4/.

Since the primary aim was to test the method of calculation under unfavorable meteorological conditions, only the axial concentrations were taken into account in the analysis. The axial concentrations were defined here as the maximum values of the concentration at a fixed distance from the plant at a given moment.

At the beginning of the process, all the values of the axial concentrations were reduced to unit total discharge of the given contaminant (1 g/sec) of the plant as a whole, on the basis of the values of the sum of the total discharges from high and low sources.

The methods for evaluating the total discharges have been discussed in /2/. It was assumed in /2/ that about 60% of the carbon disulfide consumed in xanthating and 85% of the hydrogen sulfide formed were discharged through the high stacks. The efficiency of gas scrubbing was taken into account when calculating the high discharges in Balakovo. The low discharges of hydrogen sulfide and carbon disulfide were taken as 12 and 4%, respectively. For Cherkassy, the degree of centralization of the discharges of hydrogen sulfide was much lower than for the other plants.

The Balakovo expedition of 1968 also produced data for detailed calculations from high stacks and about 300 low sources /2/.

By standardizing the concentrations it was possible to approximately compare the results of the different expeditions. Obviously, they are not completely comparable, since the meteorological conditions and the ratio between high and low discharges differed considerably on each expedition.

A small number of cases, corresponding to disconnection of the hydrogen sulfide gas scrubbing for scientific purposes, were not considered. The materials of each expedition were in general sufficiently homogeneous, since the evolution of carbon disulfide during xanthation of cellulose, representing the most important discharge into the atmosphere, the temperature and volume of the gassed air, varied negligibly from day to day. The fact that sampling was conducted in daytime only contributed to the homogeneity of the results.

The meteorological conditions during the two Balakovo expeditions were fairly similar /3/. By standardizing the concentrations for the total discharge, the small technological differences between expeditions over a period of roughly one year were eliminated to a high extent. Accordingly, in this paper we analyze results obtained in the two Balakova expeditions together, and this leads to an increased statistical reliability of the results.

Special graphs were plotted for each distance from the plant center, called the range. These graphs represent the standardized concentrations of the given contaminant as a function of the wind velocity at the wind vane level. The curve enveloping the bulk of the points from above was drawn on these graphs, so that sharply deviating concentrations, due usually to crude errors in the chemical analysis or rare anomalies of the meteorological and technological conditions, were not taken into consideration. First, the results for the ranges of 1—7 km were analyzed, since about 80% of the samples were taken in these ranges. The graphs with the enveloping curves were plotted for each expedition separately, and for the two Balakovo expeditions together (Figure 1).

The maximum value of the concentration on the enveloping curve was taken as the maximum concentration for that range. From the maximum concentrations for each range, the maximum value of c_M, corresponding to the experimentally determined dangerous wind speed u_M, was selected.

When similar curves were examined, it was necessary to bear in mind that as a rule calm weather and strong winds are very rare for convective conditions /1/. Therefore, the statistical confidence limit was lower at the extreme portions of the enveloping curves.

For a number of reasons it appeared that the results of the comparison of theoretical and experimental data were the most reliable for the concentrations of carbon disulfide. Its total discharges are estimated most accurately, since they are close to the amount of carbon disulfide used in the manufacture: hydrogen sulfide* is not a secondary product appearing at certain stages of the manufacturing process. Besides, the ratio of the carbon disulfide concentrations on the jet axis to the sensitivity of the methods of chemical analysis used was greater than the value of this ratio for hydrogen sulfide. The level of the hydrogen sulfide content in the air was frequently below the sensitivity of the chemical methods of analysis, due to the presence of a large number of so-called zero samples. And, finally, carbon disulfide is more stable under atmospheric conditions.

The results of the processing of the experimental data were compared with the concentrations calculated by the method of the

---

* [Probably carbon disulfide.]

Voeikov Main Geophysical Observatory /1/, which were also
standardized with reference to the total discharge of the corresponding component.   In the formulas for unfavorable climatic
conditions of the given region, the coefficient A was taken as equal
to 160 for Cherkassy and Balakovo and to 200 for Krasnoyarsk /1/.
The parameters of the high and low sources have been briefly
described in /2/. The computed air temperatures T_a were taken as
25° for Cherkassy and Balakovo and 0° for Krasnoyarsk, according
to the conditions of the expeditions.

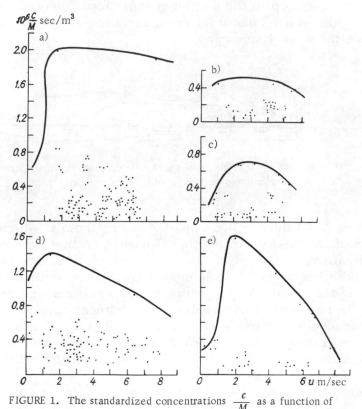

FIGURE 1. The standardized concentrations $\dfrac{c}{M}$ as a function of
the wind velocity u at a range of 2 km.
Cherkassy, carbon disulfide (a); Krasnoyarsk: carbon disulfide (b)
and hydrogen sulfide (c); Balakovo, 1967–1968: carbon disulfide (d)
and hydrogen sulfide (e).

Figure 2 shows the results of calculations of the carbon disulfide
concentration field under the jet in Balakovo during a westerly wind
of 3 m/sec, and the values of the discharge parameters obtained in
direct measurements carried out in 1968.   The origin of the

coordinates is located at the center of location of the high stacks, and the x axis is oriented in the direction of the wind.   The isolines of the standardized concentrations, drawn through $1 \sec/m^3$, were plotted from calculations for the x axis and two lines parallel to it: $y = \pm 200$ m, and also for the points corresponding to the highest concentrations from individual sources.   The latter, as shown in /3/, should ensure a sufficiently complete determination of the local maxima of the concentration field.

The location of the largest sources was taken into account in detail in the calculations, while similar small sources from a single plant were reduced to the center of their location (and not to the center of the whole industrial area, as in the simplified calculations based on the total discharges).

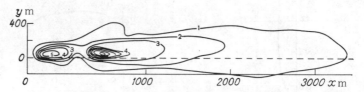

FIGURE 2. Distribution of the calculated concentration $(\sec/m^3)$ of carbon disulfide under the jet of the Balakovo rayon plant.

Figure 2 is fairly typical, and therefore from it we can draw a number of conclusions of a general nature on the distribution of concentrations in the jet of viscose rayon plants.   Such conclusions were taken into account when organizing the experimental studies, analyzing the results of the actual concentrations, and comparing them with the calculated results.   The highest concentrations corresponded to the industrial area and the adjoining regions, where the concentrations from the low sources were maximum.   The concentrations in the industrial area were 3—5 times the level of air contamination at a range of 2 km.

The calculations showed the complex structure of the concentration field in the industrial area and near it.   A detailed investigation would require a large number of sampling stations.   Such an investigation was not the aim of the expeditions, and therefore the number of samples taken at the range of 0.5—1 km was relatively small.   The concentrations decreased far from the plant, and the jet became increasingly homogeneous, as followed from the calculations and measurements.   As a result, it was possible to carry out studies under the jet with a relatively small number of sampling stations /2/.

Detailed calculations were carried out for Balakovo for several wind directions; they showed that differences in the concentrations were observed mainly in the industrial area. Starting at distances of 1—2 km from its center, the level of air contamination was relatively constant. Accordingly, detailed calculations (Figure 2) were carried out mainly for a westerly wind. Such calculations for both carbon disulfide and hydrogen sulfide were made for wind velocities of 0.5, 1, 2, 3, 4, and 5 m/sec. The dependence of the concentrations on the wind velocity in the given range of its variation was greatest in the industrial area (up to 2—3 times); the maximum level of contamination was observed for weak winds. At distances of 2 km and above the dependence of the concentrations on the wind velocity was found to be very slight.

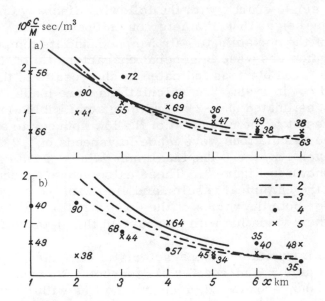

FIGURE 3. Calculated and measured concentrations of carbon disulfide (a) and hydrogen sulfide (b):

1) according to measurements in stack, 1968; 2) according to the balance, 1967; 3) according to the balance, 1968; maximum concentrations at the given range in 1967 (4) and 1968 (5); the figures indicate the number of axial samples on which the determination of the axial concentrations is based.

Figure 3 shows the maximum concentrations in Balakovo as a function of the distance x. The unbroken lines give the results of calculations according to the complete scheme for 1968, using the values of the standardized concentrations for the dangerous wind speed for each range.

For the same ranges values of the maximum measured concentrations for 1967—1968 are given together with the number of axial concentrations on the basis of which they were determined.

The agreement between the theoretical and experimental results is satisfactory.

Figure 3 also shows the results of the approximate calculations based on the total discharges of carbon disulfide and hydrogen sulfide separately for 1967 and 1968, determined from the specific balance of these components in the plants, and allowing for the efficiency of gas scrubbing of the high discharges /2/. The method of of calculation is described in /1/. The temperature and velocity of discharge of the contaminated air for the high stacks were taken as in the detailed calculations. All the sources were reduced to the center of location of the high stacks. The low sources were replaced by a single source, with the following discharge parameters: source height 15 m, diameter of its opening 1 m, velocity of discharge of the contaminated air 5 m/sec, and its temperature 25°. Such parameters were in general characteristic of the low sources of the given plant, as indicated by the results of the investigation carried out in 1968. The calculations were made for a distance $x \geqslant x_p$, estimated at 1—2 km, with an admissible error of 10—20% as a result of the reduction of the low sources to a single point /1/. The calculations were made for speeds of 1, 2, 3, 4, 5 m/sec, for each 1 m/sec. The maximum values of the concentrations are shown in Figure 3. These calculations agreed satisfactorily with the empirical results, and also with the more accurate calculations, for both the values of the concentrations and the character of their variation with increase in the distance from the plant.

In Balakovo in summer the characteristic computed dangerous wind velocity was 0.5—1 m/sec, for all sources. Naturally, the weighted mean dangerous wind velocity also lies within the same limits /1/. Therefore, at these wind velocities, the highest calculated concentrations in the industrial area and near it, i. e., at the places of maximum concentrations from the main sources, were observed. At larger distances the dependence of the calculated maximum concentrations on the wind velocity was not very marked. The deviation of the maximum concentration from its averaged value in the given range of wind velocities did not usually exceed 20—30%. A roughly similar dependence on the wind velocity was characteristic of the experimentally found concentrations as well. With a certain error we can speak of a wide range of dangerous wind velocities.

It follows in particular that when designing plants it is sufficient to calculate the weighted mean dangerous wind velocity for regions situated several kilometers from the plant area.

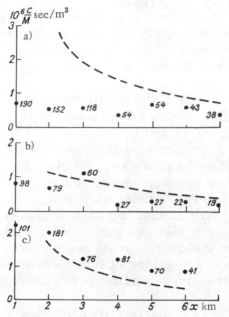

FIGURE 4. Calculated and measured concentrations in Krasnoyarsk for carbon disulfide (a) and hydrogen sulfide (b), and in Cherkassy for carbon disulfide (c). The points indicate the maximum concentrations at the given range, and the figures the number of axial samples on which their determination was based.

The main distinctive characteristic of the data of the Krasnoyarsk expedition was that they corresponded to the conditions of the cold season. The number of samples taken was relatively small, and as a result only a tentative comparison of the calculations with the experimental data was possible. In the calculations the specific total amount of carbon disulfide and hydrogen sulfide reaching the high and low stacks corresponded to the same specific balance as in Balakovo. This agreed satisfactorily with the measurements in the stacks for high discharges. The only difference was the absence of gas scrubbing in Krasnoyarsk, and so all the discharges reaching the stacks entered the atmosphere. The calculations were conducted according to the approximate procedure of /1/, with the height of the low sources, the diameter of their opening, the temperature, and velocity of discharge of the contaminated air, the same as for Balakovo. The temperature of the discharges was 25° or above and thus there was a temperature gradient of several tens of degrees in winter between the discharges and the surrounding air. This caused an increase in the dangerous wind velocities on this expedition. For high sources they were approximately 3 m/sec, for low sources 1—1.5 m/sec, the weighted mean dangerous wind velocity was about 2 m/sec.

The calculations based on the specific balance of carbon disulfide and hydrogen sulfide can be compared with the results of the observations from Figure 4a and b.

The standardized calculated concentrations of carbon disulfide were found to be roughly the same as for Balakovo. For hydrogen sulfide, in the absence of gas scrubbing, the low discharges contained relatively less, and the level of calculated standardized concentrations was roughly a half of that for carbon disulfide.

On the whole, in spite of the approximate methods for evaluating low discharges and the relatively small number of samples taken, we can conclude that the conditions of turbulent diffusion of contaminants under winter-spring conditions are less dangerous than in summer.

The number of samples of carbon disulfide and hydrogen sulfide taken in Cherkassy was roughly equal to the number taken on each Balakovo expedition. Similar weather conditions prevailed on all the summer expeditions. However, on the Charkassy expedition, which was the first and was experimental in several respects, it was not possible to measure the total discharges of carbon disulfide and hydrogen sulfide by means of the high and low discharge sources. As a result, the only possible means of estimating these discharges was by using the specific balance of carbon disulfide and hydrogen sulfide. The method of chemical analysis of the samples also differed from that used in later expeditions. In particular, the hydrogen sulfide samples were not scrubbed from sulfur dioxide.

Figure 4, c gives the results of a comparison of the experimental and calculated results for carbon disulfide. In a zone of 1—3 km, the agreement must be considered as satisfactory. Farther away the calculated concentrations decreased somewhat more rapidly than the actual values, possibly because of the sewage sedimentation basins near the plant.

The level of the observed concentrations of hydrogen sulfide corresponded to 20% of the low discharges of this component.

According to the calculations for Cherkassy, the most dangerous wind velocities in the plant area were 0.5—1 m/sec. At distances of 1—5 km, the most frequently observed wind velocities of 2—3 m/sec were roughly equally dangerous, as shown in Figure 1, a. At large distances, both the calculations and the measurements indicate that the concentrations increase somewhat with increase in the wind velocity.

We shall summarize our results.

The calculations according to the method given in /1/ have been satisfactorily confirmed by experiment. The calculations based on the specific balance of carbon disulfide and hydrogen sulfide gave positive results of roughly the same order of accuracy as the calculations based on direct measurements of the discharges from the sources.

It was shown that low discharges contribute greatly to the total concentrations up to considerable distances.

The highest and most stable level of contamination of the ground air layer was observed in summer under convective weather conditions. This agrees satisfactorily with the results of theoretical studies.

It was found that calculations of the concentrations based on the weighted mean dangerous wind velocity gave satisfactory results for viscose rayon plants.

The investigations in Balakovo confirmed the high efficiency of the hydrogen sulfide purification used there. We should note, however, that in the presence of a considerable volume of low discharges it may prove inexpedient to purify the gases entering the stack. Therefore, when designing and reorganizing plants, particular attention must be paid to a reduction of such discharges of hydrogen sulfide and carbon disulfide into the atmosphere.

It is necessary to continue studies for a further refinement of the specific balance of carbon disulfide and hydrogen sulfide. This will lead to a higher accuracy of the quantitative evaluations of the efficiency of the different sets of measures taken to protect the air basin.

The assessments of Giproiv of the values of the discharges of carbon disulfide and hydrogen sulfide in the sewage sedimentation basins during a study of the specific balance of these gases in the plants showed that these discharges may represent about 10% of the total amount of carbon disulfide consumed and of hydrogen sulfide formed. Although the extent of neutralization of these components in the sedimentation basins has not been sufficiently studied, there are reasons for assuming that under certain conditions their contribution to air contamination may be considerable. Therefore, when designing or reorganizing viscose rayon plants, we must aim at the highest possible return of the discharges in the sewage to gas scrubbing, followed by centralization of the air with sufficient concentrations in the high stacks.

Bibliography

1. Berlyand, M. E. and R. I. Onikul. K obobshcheniyu teorii rasseivaniya promyshlennykh vybrosov v atmosferu (Generalization of the Theory of the Dispersion of Industrial Discharges in the Atmosphere). See this Collection.
2. Vdovin, B. I. et al. Eksperimental'nye issledovaniya rasseivaniya v atmosfere kholodnykh ventilyatsionnykh vybrosov predpriyatii iskusstvennogo volokna (Experimental Studies on the Dispersion in the Atmosphere of Cold Ventilation Discharges of Rayon Plants). See this Collection.
3. Gracheva, I. G. et al. K raschetu zagryazneniya atmosfery ot mnogikh istochnikov (Calculation of the Contamination of the Atmosphere from Many Sources). — Trudy GGO, No. 238. 1969.
4. Onikul, R. I. et al. Rezul'taty analiza eksperimental'nykh dannykh, kharakterizuyushchikh raspredelenie atmosfernykh zagryaznenii vblizi teplovykh elektostantsii (Results of an Analysis of Experimental Data Characterizing the Distribution of Atmospheric Contaminations near Thermal Power Plants). — Trudy GGO, No. 172, pp. 23—34. 1965.
5. Ukazaniya po raschetu rasseivaniya v atmosfere vrednykh veshchestv (pyli i sernistogo gaza), soderzhashchikhsya v vybrosakh promyshlennykh predpriyatii (Instructions for Calculating the Dispersion of Contaminants (Dust and Sulfur Dioxide) Contained in the Discharges of Plants into the Atmosphere). CH-369-67. Leningrad, Gidrometeoizdat. 1967.

STUDY OF THE PROPAGATION OF CONTAMINANTS IN THE REGION OF HIGH DISCHARGE SOURCES

V. A. Belugina, N. S. Burenin,
P. I. Velikaya, B. B. Goroshko

The Voeikov Main Geophysical Observatory has carried out experimental studies in the region of thermal power plants with discharge heights of 120 m (the Shchekino and Cherepet power plants) and 180 m (the Moldavian power plant) /5, 6, 7, 9/. By means of these results the theoretical calculations of the field of concentrations made by Berlyand /2, 3, 10/ could be compared with the experimental results. A satisfactory agreement was established, and the procedures for calculating the concentrations fields of dust and sulfur dioxide generated by industrial plants /1, 10/ were approved.

It is very important to test the applicability of the method of calculation to higher sources, since today 250 m high stacks are used in many power plants, and even higher stacks are being designed. The problem, therefore, consisted in obtaining experimental data on the field of concentrations in the region of a power plant with a discharge height of 250 m. The Krivorozhye power plant was selected; this plant has two 250 m high stacks, and one 180 m high stack. The studies were conducted in June—July by the staff of the Ukrainian expedition division of the Main Geophysical Observatory.

The program of work included the conduction of gradient observations of the wind speed and direction, and also of the air temperature and humidity at heights of 0.5 and 2.0 m. The concentration field was measured under the jet at 5—6 ranges simultaneously, with periodical changes at distances from the source of 1, 2, 3, 4, 5, 6, 7, 8, 10 and 15 km. Sampling was conducted in the early morning, daytime, and evening, to include different meteorological conditions. A distinctive feature of the studies was that this was the first time that samples were taken not only of sulfur dioxide and dust, but also of nitrogen oxides and carbon monoxide. Also, two parallel samples were taken of the main components, sulfur dioxide and dust. The samples were analyzed by the standard procedure used in the Hydrometeorological Service.

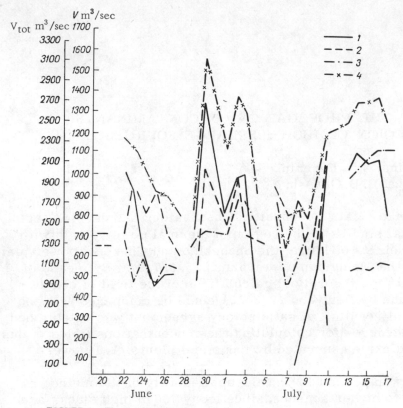

FIGURE 1. Fluctuations in the volume of smoke gases discharged from power plant stacks:

1) stack height 180 m; 2 and 3) stack height 250 m; 4) total discharge from three stacks.

The total volume of gases discharged from each stack, their temperature, and the amount of sulfur dioxide and dust discharged into the atmosphere per unit time, were measured directly at the power plant. In the gas conduits behind the filters, samples were taken for chemical analysis to determine the amounts of sulfur dioxide and nitrogen oxides.

The operation of the source was not stable during the study period, that is to say, the volume of discharged gases varied over wide limits. This is seen from Figure 1, which shows the variation in the volume of gases discharged into the atmosphere from each stack separately and all of them together. Simultaneously, there were variations in the discharge of harmful substances. Accordingly, when calculating the field of concentrations, only the maximum discharge values were taken into account, since it was these which led to the most dangerous conditions of contamination of the atmosphere. The concentrations measured under the jet were processed by the method suggested by Berlyand /8/, which consisted in plotting on a graph all the concentrations obtained for equal wind speeds and in drawing an enveloping curve through the maximum values.

The field of concentrations was calculated according to the "Instructions for calculating the dispersion in the atmosphere of harmful substances (dust and sulfur dioxide) contained in the discharges of industrial plants" /10/.

The experimental and theoretical curves of the distribution of concentrations as a function of the distance from the source agree satisfactorily, as can be seen from Table 1, which shows the maximum concentrations of dust and sulfur dioxide at different distances from the source, divided by the maximum concentration at 5 km (q/q_{max}).

TABLE 1.

Component	Data	Distance from source (km)						
		1	2	3	5	7	10	15
Dust	Calculated	0.3	0.7	0.9	1.0	0.7	0.6	0.4
	Experimental ...	0.5	0.8	0.9	1.0	0.9	0.8	0.7
Sulfur dioxide	Calculated	0.4	0.7	0.9	1.0	0.9	0.7	0.5
	Experimental	0.6	0.7	0.9	1.0	0.9	0.7	0.6

We should note that the actual concentrations are somewhat higher than the calculated. This difference is 7% at the maximum. Such a discrepancy is natural, in view of the error in the determination of the fuel consumption, and the chemical analysis of the fuel and samples. Single values of the maximum concentrations were higher than the enveloping curves. As a percentage, they represent a negligible part of the total number of samples taken: 1.4% of the samples at 1−4 km, 1% at 10−15 km, 0.03% at the maximum, were above the experimental curve. A somewhat larger discrepancy is observed at near ranges, due to the influence of low (from 30 m high stacks) and disorganized (from gas ducts, coal depots, etc.) discharges, which were not taken into account in the calculation. The influence of the disorganized discharges decreases with distance from the source, and becomes minimum when the calculated concentrations are maximum.

The calculated and actual dust concentrations agree less. The differences are 20% at the maximum, and increase at greater

distances.   Here, however, we must take into account that the weight
method of determination does not exclude the appearance of natural
dust, which may appreciably increase the concentration.   It is also
possible that the assumed 95% efficiency of the electric filters was
considerably overestimated, since the filters had not been adjusted
for a long time.   As a result, calculated concentrations were lower
than the experimental values.

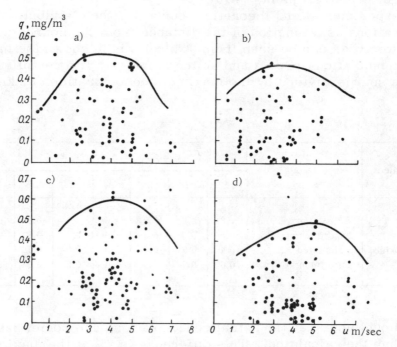

FIGURE 2.   Distribution of the concentration of sulfur dioxide as a
function of the wind velocity at different distances from the discharge source:

a) 1 km; b) 3 km; c) 5 km; d) 8 km.

Figure 2 shows the distribution of concentrations as a function
of the wind velocity at distances of 1, 3, 5, and 8 km.   The wind
velocity was calculated at the weather vane height $(h_v = 10 \text{ m})$, on the
basis of measurements at a height of 2 m, assuming a logarithmic
distribution.   It follows from the figures that the concentrations
increase with increase in the wind velocity, pass through a maximum, and then decrease.   There is thus some dangerous velocity
at which maximum concentrations are observed.   For the source

studied, according to Figure 2, $u_M = 4-5$ m/sec. The dangerous velocity was calculated from the formula

$$u_M = 0.65 \sqrt[3]{\frac{V \Delta T}{H}} ,$$

where V is the volume of the discharged gas-air mixture (m^3/sec), ΔT the difference between the temperature of the discharged gas-air mixture and the temperature of the surrounding atmospheric air, H the height of the discharge source. The calculations by means of this formula gave $u_M \approx 5$ m/sec. There is thus satisfactory agreement between the experimental and calculated values of the dangerous wind velocity.

The concentrations of nitrogen oxides were measured for the first time in the power plant area. An analysis of the samples taken showed that nitrogen oxides are contained in the discharged gases. Their maximum concentrations are approximately $1/7$ of those of sulfur dioxide.

Bibliography

1. Vremennaya metodika raschetov rasseivaniya v atmosfere vybrosov (zoly i sernistogo gaza) iz dymovykh trub elektro-stantsii (Provisional Method for Calculating the Dispersion in the Atmosphere of Discharges (Sols and Sulfur Dioxide) from the Stacks of Power Plants). — Trudy GGO, No. 172, pp. 205—212. 1965.

2. Berlyand, M. E. et al. Chislennoe issledovanie atmosfernoi diffuzii pri normal'nykh i anomal'nykh usloviyakh strati-fikatsii (Numerical Investigation of Atmospheric Diffusion under Normal and Anomalous Stratification Conditions). — Trudy GGO, No. 158, pp. 22—32. 1964.

3. Berlyand, M. E., E. L. Genikhovich, and R. I. Onikul. O raschete zagryazneniya atmosfery vybrosami iz dymo-vykh trub elektrostantsii (Calculation of the Contamination of the Atmosphere by Discharges from the Stacks of Power Plants). — Trudy GGO, No. 158, pp. 3—21. 1964.

4. Berlyand, M. E. and R. I. Onikul. Fizicheskie osnovy rascheta rasseivaniya v atmosfere promyshlennykh vybrosov (Physical Principles for Calculating the Dispersion of Industrial Discharges in the Atmosphere). — Trudy GGO, No. 234, pp. 3—27. 1968.

5. Gil'denskiol'd, R.S. et al. Rezul'taty eksperimental'nykh issledovanii zagryazneniya atmosfery v raione Moldavskoi GRES (Results of Experimental Studies of Air Pollution in the Region of the Moldavian Power Plant). – Trudy GGO, No. 207, pp. 65–68. 1968.

6. Goroshko, B.B. Postanovka eksperimental'nykh rabot po izucheniyu rasprostraneniya vrednykh primesei ot moshchnykh istochnikov (Organization of Experimental Studies on the Propagation of Contaminants from Powerful Sources). – Trudy GGO, No. 234, pp. 109–115. 1968.

7. Goroshko, B.B. et al. Meteorologicheskie nablyudeniya pri issledovanii promyshlennykh zagryaznenii prizemnogo sloya vozdukha (Meteorological Observations during a Study of Industrial Contaminations in the Ground Air Layer). – Trudy GGO, No. 138, pp. 18–30. 1963.

8. Onikul, R.I. et al. Rezul'taty analiza eksperimental'nykh dannykh, kharakterizuyushchikh raspredelenie atmosfernykh zagryaznenii vblizi teplovykh eletrostantsii (Results of an Analysis of Experimental Data on the Distribution of Atmospheric Contaminations near Thermal Power Plants). – Trudy GGO, No. 172, pp. 23–34. 1965.

9. Rikhter, B.V., R.S.Gil'denskiol'd, and V.M.Styazhkin. Raspredelenie prizemnykh kontsentratsii sernistogo gaza i zoly v zone teplovoi elektrostantsii (Distribution of Ground Concentrations of Sulfur Dioxide and Sols in the Zone of a Thermal Power Plant). – Trudy GGO, No. 158, pp. 84–87. 1964.

10. Ukazaniya po raschetu rasseivaniya v atmosfere vrednykh veshchestv (pyli i sernistogo gaza), soderzhashchikhsya v vybrosakh promyshlennykh predpriyatii (Instructions for Calculating the Dispersion in the Atmosphere of Contaminants (Dust and Sulfur Dioxide) Contained in the Discharges of Industrial Plants). CH-369-67. Leningrad, Gidrometeoizdat. 1967.

STEREOPHOTOGRAMMETRIC INVESTIGATION OF THE AIR FLOW IN THE BOUNDARY LAYER OF THE ATMOSPHERE ABOVE A HILL

V. S. Eliseev

Introduction

Ground stereophotogrammetric surveys have recently been widely applied in the national economy, because of their many specific advantages over other methods of investigation, namely: the possibility of highly accurate measurements of the coordinates of the object in three-dimensional space, the possibility of recording transient processes, and of splitting complex ones into separate phases, etc.

Stereophotogrammetric methods are used in geophysical investigations as well. Stereophotogrammetric surveys have been used to study disturbances /8/, and also problems of high-atmosphere physics: they form one of the basic methods for studying air streams and diffusion characteristics at heights of 50−200 km /17/.

This method can also be successfully used for studying wind structure and turbulent characteristics in the boundary layer of the atmosphere.

Pilot balloons are today the standard means for studying the wind profile in the lower layers of the atmosphere. However, with these we can determine only the integral wind velocity in the successive layers of the atmosphere. It is difficult to particularize the wind profile thus obtained, because of the difficulty of following the displacement of the balloon, and the inadequate reaction of the balloon to different types of air motion, in particular, as the wind scale decreases /19/.

Another approach to the solution of the above problem consists in using a number of long-range anemometers attached to the balloon cable. We must take into account not only some difficulties in its servicing, but also the practical impossibility of launching several balloons if we need to know the spatial structure of the air streams.

Recently, high masts with wind sensors mounted on them have been used in meteorological practice. An example of such a mast in the USSR is the 300 m high meteorological tower at Obninsk. While high masts are at present one of the best means for determining the vertical wind profile in the lower 300—500 m layer of the atmosphere, they also suffer from several drawbacks: lack of mobility, high cost of construction and maintenance, aviation hazards.

Therefore, in many countries attempts have been made to develop a method for determining the wind profile without these drawbacks. Such a method is photography of the drift and deformation of a vertical smoke jet created by a falling smoke rocket, or of a smoke trail left by a smoke bomb.

One variant of this method was used in 1952 by the Cambridge Aviation Research Center together with the Directorate of Geophysical Research /16/. A smoke bomb was ejected from a helicopter at an altitude of 1.5 km, and the smoke trail left by it was photographed. However, to create an unbroken smoke trail involved considerable technical difficulties. Therefore, in later experiments smoke grenades were attached to a steel cable, which was raised by means of a helicopter. The grenades were ignited by an electric pulse, and a smoke cloud was formed.

The first attempt to raise a 200 m steel cable was made in February 1953. Considerable difficulties were encountered, since when the cable hung freely its lower end was deflected by 600 m in the direction of the wind, and the cable was frequently damaged on contact with the ground. It took some time to repair the long and unwieldy cable after each experiment, and thus it was impossible to carry out tests according to a given schedule. At the same time, the tests showed that the visibility and permanence of the smoke clouds was very good. The coordinates of the smoke clouds were determined graphically by means of special calibration nomograms.

Recently, studies have been carried out in East Germany, with smoke trails created by smoke aircraft bombs. The smoke trails formed by smoke mortar shells were photographed by means of two aircraft cameras. The "normal case" of exposure was used, and the negatives were processed on a high-accuracy stereocomparator /14/.

The last decade is characterized by the increased interest of scientists in the problems of air pollution. The procedures developed for calculating the propagation of contaminants from powerful industrial sources are applicable to relatively flat areas /2,18, et al/, while many of the polluting sources are located in regions of complex relief, in which the spatial distribution of the wind follows its own laws.

There are numerous theoretical /7, 10, 11/ and microclimatic experimental /9, 12/ studies on the influence of complex relief shapes on the air flow. The results of microclimatic studies correspond to the ground air layer up to heights of 2—4 m, while the theoretical studies are connected with a certain stylization of the formulation of the problem. A cycle of theoretical /3, 4/ and experimental /6/ studies on the influence of small-scale complex relief shapes on the flow structure and diffusion of contaminants was conducted at the Voeikov Main Geophysical Observatory. One of these studies consisted in developing methods for a ground stereophotogrammetric survey for determining the eddy coefficients of diffusion from the visible outlines of the smoke jets, and establishing the wind field structure laws above irregular and complex relief shapes.

Implementation of the method

The methods of ground stereophotogrammetric survey can be used to investigate the structure and diffusion characteristics of the air flow in the boundary layer of the atmosphere, if the following four types of equipment are available: safe generators of smoke trails, a stereophotogrammetric device, an automatic apparatus for processing stereopairs, an electronic computer for solving the algorithm of transformation of the coordinates of the photographed object, and calculating the necessary parameters of the airflow structure.

Since we selected the analytical method (as the most accurate one) for solving the equations connecting the coordinates of the investigated targets and the coordinates of their images, together with the Moscow Institute of Engineers of Geodesy, Aerial Survey, and Cartography, we developed universal equipment for the ground stereophotogrammetric survey of smoke trails, and it became possible to use the "general case" of exposure /5/.

The photographic equipment was based on the Soviet-made aerial cameras AFA-41/20, together with high-accuracy optical theodolites. Aerial cameras were preferable to the usual phototheodolites because of the error in the determination of the coordinates caused by the small photographic plates used in the phototheodolite method. Besides, the minimum discreteness of the aerial camera photography is 2.2 sec instead of 1—2 min. Photographs are taken on a black-and-white or color reversible film with frame size 19 × 19 cm. To eliminate errors due to film deformation, the film was pressed on glass, and during the photography coordinate crosses were printed over the whole image field. Figure 1 shows the possible three-dimensional scale of the photography with this equipment as a function of Y_{max}. This scale is used for the following characteristics:

Y_{max} (m)	400	800	1,200	1,600	2,000	2,400	2,800	3,200	3,600	4,000
Y_{min} (m)	80	160	240	320	400	480	560	640	720	800
Depth of photography	320	640	960	1,280	1,600	1,920	2,240	2,560	2,880	3,200
Horizontal scope (m)										
max	360	720	1,080	1,440	1,800	2,160	2,520	2,880	3,240	3,600
min	72	144	216	288	360	432	504	576	648	720
Area of photography (km^2)	0.069	0.276	0.622	1.106	1.728	2.488	3.387	4.424	5.599	6.912
h_{max} (m)	180	360	540	720	900	1,080	1,260	1,440	1,620	1,800
h_{min} (m)	36	72	108	144	180	216	252	288	324	360

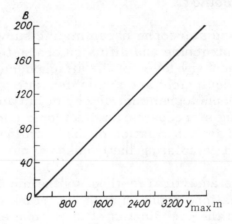

To ensure the efficiency of the analytical method for processing the stereopairs, an automated "Stekometr" precision stereocomparator and a digital computer are used instead of the standard stereocomparator. The rate of coordinate reading is thus increased by a factor of 5 to 7, and operator errors are eliminated. The mean error of measurement of the coordinates is $\pm 2 \cdot 10^{-3}$ mm. A teletype is used to print the results directly in decimal notation, and to punch them on a paper tape which is then fed to a M-220 computer for processing. Thus, accurate calculations can be made for a very large number of points.

Standard pyrotechnical devices, generating vertical orange-colored strips starting at ground level, were used to create the

smoke trails.   These devices are safe to use and relatively inexpen-
sive.   The smoke trail persists for one minute, and the color
characteristic creates not only a brightness contrast but also a
color contrast, which makes it possible to obtain a satisfactory
image on the negative for any background.

Determination of the wind speed from stereophotographs and error analysis

We shall consider several questions on the accuracy of calculation of the air speed from stereophotograms of vertical smoke
trails.
Let X_{0i}, Y_{0i}, Z_{0i} be the coordinates of singular points of the smoke
trail obtained at the initial moment t_0.   The values of these coordinates in the "normal case" of exposure are determined from the
formulas:

$$X_{0i} = B x_{0i} p_{0i}^{-1};\tag{1}$$

$$Y_{0i} = B f_k p_{0i}^{-1};\tag{2}$$

$$Z_{0i} = B z_{0i} p_{0i}^{-1}.\tag{3}$$

As a result of the shift in the singular points in the wind field,
their coordinates acquire new values X_{1i}, Y_{1i}, Z_{1i}.   Formulas (1)–(3)
for the numerical values of the coordinates can then be written in
the form:

$$X_{1i} = B x_{1i} p_{1i}^{-1};\tag{4}$$

$$Y_{1i} = B f_k p_{1i}^{-1};\tag{5}$$

$$Z_{1i} = B z_{1i} p_{1i}^{-1}.\tag{6}$$

Therefore, the coordinates of the path traveled by the singular points
as a result of their drift in the wind, neglecting the small term
$(p_{0i} \Delta p)$, will be

$$\Delta X = B \{ [(x_{1i} - x_{0i}) p_{0i} - x_{0i} \Delta p]\, p_{0i}^{-2} \};\tag{7}$$

$$\Delta Y = - B f_k p_{0i}^{-2} \Delta p;\tag{8}$$

$$\Delta Z = B \{ [(z_{1i} - z_{0i}) p_{0i} - z_{0i} \Delta p] p_{0i}^{-2} \}.\tag{9}$$

Thus, the wind velocity and its X, Y, Z components are determined
from measurements of two stereopairs, obtained at two moments of

time separated by a specified time interval. Since, from (7)–(9)

$$\Delta X = \Phi_1(B, x_{1i}, x_{0i}, p_{0i}, \Delta p),$$
$$\Delta Y = \Phi_2(B, f_k, p_{0i}, \Delta p),$$
$$\Delta Z = \Phi_3(B, z_{1i}, z_{0i}, p_{0i}, \Delta p),$$
$$\Delta p = \Phi_4(p_{1i}, p_{0i}).$$

the squares of the mean errors of the displacement of the singular points of the smoke jet along the X, Y, Z axes can be written in the form

$$S_{\Delta X}^2 = \left(\frac{d\,\Delta X}{dB}\right)^2 S_B^2 + \left(\frac{d\,\Delta X}{dx_{1i}}\right)^2 S_{x_{1i}}^2 + \left(\frac{d\,\Delta X}{dx_{0i}}\right)^2 S_{x_{0i}}^2 + \tag{10}$$
$$+ \left(\frac{d\,\Delta X}{dp_{0i}}\right)^2 S_{p_{0i}}^2 + \left(\frac{d\,\Delta X}{dp_{1i}}\right)^2 S_{p_{1i}}^2; \tag{11}$$

$$S_{\Delta Y}^2 = \left(\frac{d\,\Delta Y}{dB}\right)^2 S_B^2 + \left(\frac{d\,\Delta Y}{df_k}\right)^2 S_{f_k}^2 + \left(\frac{d\,\Delta Y}{dp_{0i}}\right)^2 S_{p_{0i}}^2 + \left(\frac{d\,\Delta Y}{dp_{1i}}\right)^2 S_{p_{1i}}^2;$$

$$S_{\Delta Z}^2 = \left(\frac{d\,\Delta Z}{dB}\right)^2 S_B^2 + \left(\frac{d\,\Delta Z}{dz_{1i}}\right)^2 S_{z_{1i}}^2 + \left(\frac{d\,\Delta Z}{dz_{0i}}\right)^2 S_{z_{0i}}^2 + \left(\frac{d\,\Delta Z}{dp_{0i}}\right)^2 S_{p_{0i}}^2 +$$
$$+ \left(\frac{d\,\Delta Z}{dp_{1i}}\right)^2 S_{p_{1i}}^2. \tag{12}$$

After the values of the derivatives for equations (10)–(12) have been determined, and neglecting the small terms, we obtain the theoretical value of the mean square error in the wind speed during the time interval Δt for $S_{p_{0i}} = S_{p_{1i}} = S_p;$ $S_{x_{0i}} = S_{x_{1i}} = S_x$ and $S_{z_{0i}} = S_{z_{1i}} = S_z$ which is equal to

$$S_V = A^{1/2} \Delta t^{-1}, \tag{13}$$

where

$$A = 2B^2 p^{-4}(S_x^2 p^2 + S_p^2 x_1^2 + S_z^2 p^2 + S_p^2 z_1^2) + 2\Delta Y^2 S_p^2 \Delta p^{-2}.$$

When calculating the mean wind speed on the basis of several measurements N:

$$S_{\bar V} = A^{1/2} N^{-1/2} \Delta t^{-1}. \tag{14}$$

And, finally, the mean square error of the fluctuations in wind speed is

$$S_{V'} = A^{1/2} \Delta t^{-1}(1 + N^{-1/2})^{1/2}. \tag{15}$$

By assigning numerical values to the input parameters close to the values obtained in the given experiment, then when $B = 135$ m, $\Delta t = 3$ sec, $\Delta Y = 3$ m, $N = 5$, $x_1 = 60$ mm, and $z_1 = 40$ mm we obtain

$$S_V = 0.24 \text{ m/sec},$$
$$S_{\bar{V}} = 0.11 \text{ m/sec},$$
$$S_{V'} = 0.29 \text{ m/sec}.$$

We should note that the error may vary over the photograph field as a function of x_1 and z_1.

If we use the trails of the smoke directions for the calculations, we obtain the wind speed for sufficiently brief time intervals. However, it is frequently necessary to obtain the average wind profile over a time of the order of 10 min. We must therefore determine how many implementations of stereopairs with smoke traces are necessary to obtain reliable ten-minute average values of the wind speed.

If we assume that the distribution of the velocity components above flat country obeys Gauss' law (this type of velocity distribution is shown in /16/), we can write the probability density of the fluctuations in velocity in the form

$$F(u') = 2\,\pi^{-1/2}\,\sigma^{-1}\,\exp\left[-\frac{u'^2}{2\,\sigma^2}\right], \tag{16}$$

where the fluctuations in velocity are reckoned for the selected mean values.

The probability that the mean values of the velocity obtained from N implementations will differ from the ten-minute average by not more than α is then determined from the Laplace function

$$2\,\Theta(\varepsilon) = \frac{2}{\sqrt{2\pi}}\int_0^\varepsilon \exp\left[-\frac{\varepsilon^2}{2}\right]d\varepsilon, \tag{17}$$

where

$$\varepsilon = \alpha\,\sigma^{-1}\,N^{1/2}.$$

At the same time, Durst /13/ found that with a correlation of 0.87 for a wind velocity varying between 4 and 18 m/sec

$$\sigma\,(10 \text{ min } 05 \text{ sec}) = -0.2 + 0.14\,V.$$

For such a connection between σ and the ten-minute mean velocity of the air stream, formula (17) for investigating the above problem with the necessary number of discrete measurements will be

$$2\,\Theta(\varepsilon) = \frac{2}{\sqrt{2\pi}} \int_{0}^{\frac{\alpha N^{1/2}}{-0.2+0.14V}} \exp\left[-\frac{\varepsilon^2}{2}\right] d\,\varepsilon. \tag{18}$$

If we assume that $2\Theta(\varepsilon) = 95\%$ and $\alpha = 0.5\,\text{m/sec}$, when $V = 4\,\text{m/sec}$, then $N = 2$, and when $V = 10\,\text{m/sec}$, then $N = 23$.

The calculations correspond to the case when the distribution of the velocity components is Gaussian. This may not be true in the case of rough country, and then the problem of time averaging cannot be solved in this way. At the same time, if we consider the hill as a characteristic minimum distance on which there must be a dynamic effect of the hill itself, then $\tilde{T} = \frac{\tilde{L}}{V}$, where $\tilde{L}$ is the characteristic scale of the obstacle, i.e., the hill length, and $\tilde{T}$ is the characteristic time. Then, if the wind velocity varies from 2 to $20\,\text{m/sec}$, for $\tilde{L} = 500\,\text{m}$, the characteristic time of the averaging will vary from 4 min to 25 sec. For an average wind speed of $5\,\text{m/sec}$, $\tilde{T} \approx 2\,\text{min}$.

Experimental study of the air flow past a hill

As all the four components necessary for using the method of the stereophotogrammetric ground survey of smoke traces were available, in the summer of 1969, it was possible to study the space structure of the air flow above a hill under natural conditions in the Razdan valley. This experimental study was carried out by the staffs of the Main Geophysical Observatory (Genikhovich, Zaitsev, Ivanova, etc.), Moscow Institute of Engineers of Geodesy and Aerial Surveying and Cartography (Novakovskii, Sintsov) and BRIS, Administration of Hydrometeorological Service, Armenian SSR (Kosoyan).

In accordance with the formulated problem, a 100 m high and 500 m long hill was selected in the valley. The hill slopes were covered by scorched grass, and had a slope of 15—30°.

Above the hill (Figure 2) orange-colored vertical smoke traces were made, which were then photographed by a stereophotogrammetric installation with a base length of 135 m between the cameras.

FIGURE 2.

Photographs were taken on black-and-white and on color reversible film of type TsO-2 at intervals of 3 sec, until complete dispersion of the smoke traces. Gradient observations of the air temperature and the wind speed were carried out near the stereophotogram-metric equipment.

About 500 stereopairs were taken in all, and the photographs were taken in the morning and evening. With shorter intervals between successive stereoimages, the measurement of the coordinates in space of the movement of the traces is easier and more accurate since the diffusion-caused alteration in the smoke trace elements is negligible over a three-second interval. An almost error-free identification of these elements on two successive stereopairs is thus possible. The material was processed on the basis of the singular points at the jet boundaries and of equidistant points along the height. A comparison of the two methods for de-termining the wind profile showed their satisfactory agreement. The transformation of the stereophotogrammetric coordinates into coordinates in space, and the determination of the wind velocity and its components, were performed on a M-220 computer by a specially prepared program.

From the complex three-dimensional pattern of the actual air flow above a ridge perpendicular to it, it is possible to isolate the profile of the three components of the wind velocity and the overall wind velocity. Due to the absence of a temperature sounding, variations in the wind profile are considered from the point of view of the dynamic influence of the hill on the flow only. The influence of a ridge located at a distance of 1 km from the place where the experiment was conducted is eliminated by selecting the thickness of the boundary layer studied sufficiently small compared to the distance to the ridge $(\frac{H}{x'} = 0.1)$.

Figure 2 shows local bendings clearly, and we can thus conclude that in addition to the large disturbances there are even finer fluc-tuations in the velocity field. Such small-scale disturbances are obviously connected with the appearance of disturbances in the zone of velocity and temperature gradients.

Figure 3 gives as an example the distribution of the relative velocities u/u_0 in different sections of the hill. Here u_0 is the velocity of the air flow at a distance of 50 m from the foot of the hill. Since photographs were taken at intervals $\Delta t = 3$ sec, vertical lines on the figure represent the maximum deviations in the absolute values u/u_0 over a two-minute time interval. Due to the low-con-trast image of the lower part of the smoke trace, the minimum height at which the trace shift could be determined was 5 m.

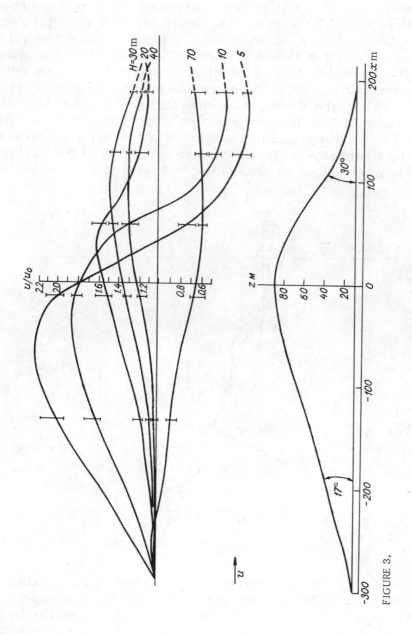

FIGURE 3.

It is seen from Figure 3 that in the front part of the hill a gradual increase in the wind speed is observed up to heights of 40—50 m. The maximum of the relative speed at heights of 5—40 m decreases from 120 to 30% of the speed in the impinging flow during its shift from $x = 250$ m ($H = 5$ m) to $x = 470$ m ($H = 40$ m). Behind the hill the flow is decelerated to 70% at a height of 5 m. The isolines u/u_0 reflect the general features of the hill configuration, but as indicated above, with a shift in the maximum against the stream at $H = 5$ m and along the stream from the hill top at greater heights.

At a height of 70 m the configuration of the isolines has a mirror image. The windward and leeward sides of the hill are characterized by a decrease in the wind speed compared to the impinging flow, with lower values of $u/u_0 = 30\%$ ($x \approx 350$ m) at the leeward side.

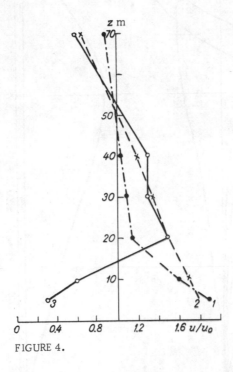

FIGURE 4.

Figure 4 represents the vertical distribution of the relative velocities u/u_0 at different sections of the hill: at a distance $x = -150$ m from the hill top (1); above the hill top (2); on the leeward side of the hill at a distance $x = 100$ m (3). The figure shows that the 50-m thick air layer behind the hill is characterized by increased wind speeds, and the 15-m thick air layer from the surface of the ground by lower speeds. The direction of the air streams in

this layer is opposite to the general direction of the stream impinging on the hill.   An elliptic vortex is formed with half-axes
$a_1 \approx 75$ m and $a_2 \approx 100$ m with wind speeds of 0.5 m/sec in its lower
part, and 2.5 m/sec in its upper part.   The highest variability in
the relative wind speed is observed at a height of 5 m from the hill
surface, with a gradual decrease in the absolute value of u/u_0 with
increase in height.   An increase in the wind speed in the front part
of the hill is also noted in /1/, where from observations of the flight
of balanced damped balloons above slopes $\alpha \approx 10°$, it was noted that
the maximum wind speeds are observed at a height of 4—6 m from
the slope surface, and are 6—8 m/sec.   According to our results,
the wind speed in the middle part of the hill at a height of 5 m was
5.5 m/sec.   But while the vertical profile u/u_0 in the middle part of
the windward slope has a maximum at a height of 5 m, drops sharply
to a height of 20 m, and then decreases more slowly to a height of
50 m, the profile at the hill top shows a continuous decrease to a
height of 50 m.
     With subsequent processing of the material it is possible to
analyze all three components of the wind velocity at different hill
sections and to clarify the principles of the structure of the air
flow in the morning and evening.

Bibliography

1. Asitashvili, A.V., K.A. Sapitskii, and Z.V. Khvelelidze.
        Izuchenie lokal'noi tsirkulyatsii vetra v raione Dusheti
        Gruz SSR (Study of the Local Wind Circulation in the
        Dusheti Region, Georgian SSR). — Meteorologiya i Gidro-
        logiya, No. 5. 1968.
2. Berlyand, M.E. and R.I. Onikul. Fizicheskie osnovy
        rascheta rasseivaniya v atmosfere promyshlennykh
        vybrosov (Physical Principles for Calculating the Dis-
        persion of Industrial Discharges in the Atmosphere). —
        Trudy GGO, No. 234, pp. 3—27. 1968.
3. Berlyand, M.E., E.L. Genikhovich, and V.K. Dem'ya-
        novich. Nekotorye aktual'nye voprosy issledovaniya
        atmosfernoi diffuzii (Some Topical Problems of the
        Investigation of Atmospheric Diffusion). — Trudy GGO,
        No. 172, pp. 3—22. 1965.
4. Berlyand, M.E., E.L. Genikhovich, and O.I. Kurenbin.
        Vliyanie rel'efa na rasprostranenie primesi ot istochnikov
        (Influence of the Relief on the Propagation of a Contaminant
        from a Source). — Trudy GGO, No. 234, pp. 28—44. 1968.

5. Bu r ov, M. I., V. S. E l i s e e v, and B. A. N o v a k o v s k i i. Stereo-
        fotogrammetricheskii metod issledovaniya atmosfernoi
        diffuzii (Stereophotogrammetric Method for Studying
        Atmospheric Diffusion). — Trudy GGO, No. 238. 1969.
6. G o r l i n, S. M. and I. M. Z r a z h e v s k i i. Izuchenie obtekaniya
        modelei rel'efa i gorodskoi zastroiki v aerodinamicheskoi
        trube (Studies of the Flow past Models of Relief and Urban
        Built-Up Areas in a Wind Tunnel). — Trudy GGO, No. 234,
        pp. 45—59. 1968.
7. D o r o d n i t s y n, A. A. Vliyanie rel'efa zemnoi poverkhnosti
        na vozdushnye techeniya (Influence of the Soil Surface
        Relief on Air Flows). — Trudy TsIP, No. 21. 1950.
8. K r y l o v, Yu. M. Spektral'nye metody issledovaniya i rascheta
        vetrovykh voln (Spectral Methods for Studying and Calcu-
        lating Wind Waves). Leningrad, Gidrometeoizdat. 1966.
9. G o l' t s b e r g, I. A. (Editor). Mikroklimat kholmistogo rel'efa
        i ego vliyanie na sel'skokhozyaistvennye kul'tury (Micro-
        climate of Undulating Contours and its Influence on Agri-
        cultural Crops). Leningrad, Gidrometeoizdat. 1962.
10. M o n i n, A. S. Model' vetra sklonov (Model of the Wind of
        Slopes). — Trudy TsIP, No. 8. 1948.
11. M u s a e l y a n, Sh. A. Volny prepyatstvii v atmosfere (Waves of
        Obstruction in the Atmosphere). Leningrad, Gidrometeoiz-
        dat. 1962.
12. S o l o m a t i n a, N. I. O vliyanii rel'efa na meteorologicheskie
        kharakteristiki v prizemnom sloe vozdukha (Influence of
        Relief on the Meteorological Characteristics in the Ground
        Air Layer). — Trudy GGO, No. 172, pp. 58—69. 1965.
13. D u r s t, C. S. Wind Speeds over Short Periods of Time. — Met.
        Mag., Vol. 89. 1960.
14. K o l b i g, J. Die Ermittlung des Windprofils bis 300 m über
        Grund durch photogrammetrische Vermessung von Rauch-
        markierungen. — Z. Met., Vol. 17. 1965.
15. L e t t a y, H. and B. D a v i d s o n. Exploring the Atmospheres
        First Mile, Vols. 1, 2. 1957.
16. L e t t a y, H. and W. S c h w e r d t f e g e r. Untersuchungen über
        atmosphärische Turbulenz und Vertikalaustausch von
        Treiballon. — Met. Z., Vol. 50. 1937.
17. L l o y d, K. and L. S h e p p a r d. Atmospheric Structure at
        130—200 km Altitude from Observations on Grenade Glow
        Clouds during 1962—1963. — Austrian J. Phys., p. 323. 1966.
18. M e a d e, P. S. Meteorological Aspects of Peaceful Uses of
        Atomic Energy. Part 1. — WMO. Tech. Note. No. 33. 1960.
19. S c o g g i n s, J. R. Aerodynamics of Spherical Balloon Wind
        Sensors. — J. Geophys. Res., Vol. 69. 1964.

WIND DIRECTION FLUCTUATIONS IN THE BOUNDARY LAYER OF THE ATMOSPHERE

A. S. Zaitsev

Wind direction fluctuations are a very important characteristic in a study of the diffusion of contaminants from different types of sources. According to theoretical investigations /1/, wind direction fluctuations are one of the characteristics determining the width of the smoke jet and the dispersion of the contaminants in a direction perpendicular to the main transfer. A number of experimental studies /2, 3, 5, 7/ showed the quantitative laws of the dependence of the scatter of the horizontal wind components on the wind velocity and the instability of the atmosphere in the ground layer. It was found that the scatter in the direction of the wind decreases on an average from 10—15 to 2—3° when the absolute value of the mean wind velocity increases from 1 to 10 m/sec, and depends on the state of stability of the atmosphere.

When studying the dispersion of contaminants from high sources (the height of the stacks of contemporary power plants reaches 200—250 m), the scatter in the wind direction over the entire boundary layer of the atmosphere must be known. Experimental studies of the boundary layer are very limited /7/, and have been carried out on special meteorological masts only.

Frequent basic pilot balloon observations were carried out during a study of the propagation of discharges from the Balakovo synthetic fiber plant in July—August 1967. The procedure used and some of the results of the analysis of these observations were published in /4/.

Three hundred and eighty seven pilot balloons in 27-day series were released during the observation period. Theodolite readings were taken every 20 sec: 10 pilot balloons at least were released during a 2-hour series. The wind characteristics were calculated on a "Ural-4" computer by the standard processing method /6/. Next, the mean wind direction at the given layer during the 2-hour period and the mean square deviation (Φ_0) were calculated. Constant (in sign) deviations in the wind direction, according to time, usually observed during the gradual reversal of the wind, were excluded from the analysis.

Before proceeding to an analysis of the results, we must consider the accuracy of the calculations of the wind direction from the results of basic pilot balloon observations. Special experiments to test the accuracy showed that the errors of the observation method for wind direction do not exceed 10° in 90% of the cases. Since these experiments were carried out under unfavorable (from the point of view of pilot balloon observations) conditions, this value can be considered as an overestimate.

The following diurnal profile of the scatter of the wind direction was found from measurements in the Balakovo plant vicinity (Table 1).

TABLE 1.

Time (hours)	Height (m)												
	50	100	200	300	400	500	600	700	800	900	1,000	1,200	1,400
10—12	17	14	14	13	14	10	11	10	10	90	8	—	—
14—16	16	16	18	14	15	20	13	16	18	17	15	11	18
7—9 and 17—19	8	10	8	10	12	11	10	9	10	10	8	8	—

Table 1 shows that in the morning and evening (the results for these two periods differed little and were combined), the fluctuations in the wind direction are not large, are relatively constant in height, and are characterized, on the average, by values of the order of 8—10°. About midday an increase in the wind direction fluctuations are observed in the lower part of the boundary layer of the atmosphere. This is apparently due to the development of thermal turbulence, and the values of Φ_0 increase by a factor of 1.5—2. Toward evening the intense turbulent exchange covers large layers. The wind speed increases over the whole boundary layer. The vertical profile becomes weakly expressed, the scatter is maximum, and the stratification of the boundary layer into separate layers is clearly visible (this is apparently due to jet flows in the boundary layer, i. e., to local acceleration of the wind speed).

Studies of the lower part of the boundary layer /7/ showed the influence of the underlying surface inhomogeneity on the scatter in wind direction, in particular, in the lowest layers. It was natural to expect that small inhomogeneities (the influence of large relief shapes is not considered here) would not appreciably affect the wind direction fluctuations in the boundary layer.

TABLE 2.

Wind direction (deg)	Height (m)										
	50	100	200	300	400	500	600	700	800	900	1,000
0–90	12	12	12	11	14	13	11	10	9	10	8
90–180	19	19	16	17	15	17	15	11	14	14	16

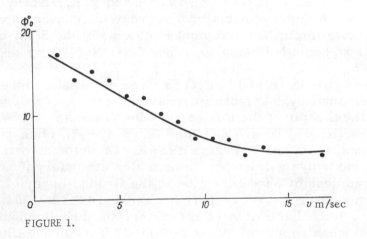

FIGURE 1.

Analysis showed that the existing dependence on the wind direction is rather connected with the overall synoptic situation during the period the experimental studies are carried out. Thus, at wind direction of 0–90° the scatter in direction is on the average 2–6° less than at directions of 90–180° (Table 2). The analysis also showed that wind directions of 0–90° were characteristic of stable synoptic processes, while directions of 90–180° characterized a change in the synoptic situations and a appreciable nonuniformity of the processes.

An analysis of the dependence of the scatter in wind direction on the wind speed showed a clear decrease in Φ_0 with increase in the wind speed (Figure 1). The dots in the figure represent the mean values. The number of cases of averaging for each range of wind speeds fluctuated between 15 and 40. A characteristic feature of the relationship shown is that decrease in the scatter stops at velocities of 12–13 m/sec. The case is similar in the ground layer also /2/, but here the decrease stopped at a velocity of 7 m/sec.

No clear dependence of Φ_0 on the stability was established. This is probably because of the limited amount of experimental data.

We can only note that superadiabatic conditions (according to the values of $\Delta T / v^2$, where ΔT is the temperature difference at the layer boundaries, and $\bar{v}$ the mean wind velocity in the layer) correspond to the highest values of the scatter of wind direction.

Bibliography

1. Berlyand, M. E., E. L. Genikhovich, and V. K. Dem'yano-
       vich.   Nekotorye aktual'nye voprosy issledovaniya atmos-
       fernoi diffuzii (Some Topical Problems of the Study of
       Atmospheric Diffusion). − Trudy GGO, No. 172, pp. 3−22.
       1965.
2. Genikhovich, E. L. and V. P. Gracheva.   Analiz dispersii
       gorizontal'nykh kolebanii napravleniya vetra (Analysis
       of the Scatter of the Horizontal Fluctuations in the Wind
       Direction). − Trudy GGO, No. 172, pp. 42−47. 1965.
3. Gracheva, V. P. and V. P. Lozhkina.   Ob ustoichivosti
       napravleniya vetra v prizemnom sloe atmosfery (The
       Steadiness of Wind Direction in the Ground Layer of the
       Atmosphere). − Trudy GGO, No. 158, pp. 41−45. 1964.
4. Zaitsev, A. S.   Issledovaniya kharakteristik turbulentnosti s
       pomoshch'yu sharov-pilotov (Study of Turbulence Charac-
       teristics by Means of Pilot Balloons). − Trudy GGO, No. 238.
       1969.
5. Ariel', N. Z.   Nekotorye rezul'taty nablyudenii za pul'satsiyami
       temperatury i napravleniya vetra (Some Results of Ob-
       servations on Fluctuations in Temperature and Wind
       Direction). − Trudy GGO, No. 107, pp. 60−65. 1961.
6. Il'in, B. M. and I. I. Chestnaya.   Obrabotka bazisnykh
       sharopilotnykh nablyudenii na EVM Ural-4 (Processing
       Basic Pilot Balloon Observations on the Ural-4
       Computer). − Trudy GGO, No. 226. 1968.
7. Munn, R. E. and A. Reimer.   Turbulence Statistics at 30 and
       200 Feet at Pinawa. − Manitoba Atm. Env., Vol. 2. 1968.

BEHAVIOR OF GROUND INVERSIONS OF THE AIR TEMPERATURE IN THE 0.5—2 m LAYER
(according to the results of gradient observations)

T. A. Ogneva, L. I. Kuznetsova

Information on the temperature inversions of the ground air layer is very important when studying questions of turbulent mixing, which determines not only the transfer of heat, moisture, and momentum in the atmosphere, but also such practically important phenomena as the intensity of dispersion of contaminants, the propagation of light electromagnetic waves, etc.

Turbulent exchange decreases strongly under inversion conditions, and it is therefore particularly important to know all the laws of the distribution of inversions in time and space. Usually, inversions of the air temperature of the ground layer are characterized by the soil-air temperature differences. However, due to the different heating mechanisms of the soil and air, it is not always possible to determine uniquely the temperature stratification in the air layer in question. It seems, therefore, advisable and interesting to describe the air temperature inversions in the 0.5—2 m layer on the basis of special gradient observations at heat-balance stations.

According to the program of heat-balance stations, measurements of the air temperature at levels of 0.5—2 m are carried out in the warm period of the year (at positive temperatures) 6 times daily (at 1, 7, 10, 13, 16, 19 hours local time), and in the cold period twice daily (at 1 and 13 hours).

The temperature is determined as the mean value of five readings of the dry thermometer of an aspiration psychrometer located at the given level in the warm period, and as the mean value of three readings in the cold period. In the cold period of the year measurements by means of an aspiration psychrometer are carried out up to the limit of the thermometer scale only, i. e., -30°. The observations are made at the sites of the meteorological stations, which are covered by natural meadow vegetation characteristic of the given geographic region. The level at which the psychrometers are set is measured from the soil surface. Data on the wind speed used

in this study were also obtained during thermal balance observations with a manual anemometer set at a height of 2 m, and exposed for 10 minutes in the observation period. The procedure is
described in the "Handbook" /1/.

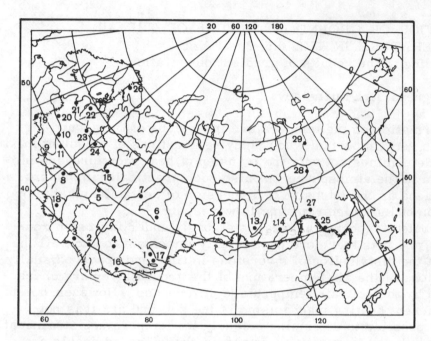

FIGURE 1. Map of the stations where observations are made for calculating
inversion frequencies.

    A list of the stations used for data processing is given in Table 1,
and their distribution over the territory is shown in Figure 1.   The
observation stations are located in typical geographical zones.
    The data of observations over a 5—7 year period (3 years in the
cold period) were used in the study.   Cases of negative differences
in the air temperature (Δt) in the layer of 0.5—2 m were picked out
when processing the experimental material.   The data were divided
into two groups according to the intensity of the inversions for Δt
between -0.1 and -0.5, and for $\Delta t < -0.5$; into three groups according
to the wind speed at the level of 2 m: <2 m/sec, 2—4 m/sec, and
>4 m/sec, and every 10° according to the air temperature.
    The stability of the sampling when averaging over different
periods was also shown by the example of observations at several
stations.   A relationship was found between the temperature inversions in the 0.5—2 m layer and the temperature difference in the

0—2 m layer, and the probability of formation of inversions in day-time in the absence of direct radiation (overcast sky with low clouds) was determined.

TABLE 1. List of the stations where observations were made for studying the inversions of the ground air layer (0.5—2 m)

No.	Name	Years of observation	Geographical zone
1	Aidarly	1963—1968	Desert
2	Akmolla	1962—1967	"
3	Beki-Beng	1962—1968	"
4	Tamdy	1961—1967	"
5	Kalmykovo	1962—1967	"
6	Tselinograd	1962—1968	Steppe
7	Rudnyi	1963—1968	"
8	Gigant	1961—1967	"
9	Askaniya-Nova	1961—1967	"
10	Borispol'	1962—1967	"
11	Poltava	1962—1967	"
12	Solyanka	1963—1968	"
13	Khomutovo	1963—1968	"
14	Chita	1964—1968	"
15	Kuibyshev	1962—1967	"
16	Dushanbe	1963—1968	Wide valleys
17	Frunze	1964—1968	" "
18	Telavi	1962—1967	" "
19	Beregovo	1961—1967	Mixed wood
20	Pinsk	1963—1968	" "
21	Riga	1962—1967	" "
22	Nikolaevskoe	1963—1967	" "
23	Smolensk	1962—1967	" "
24	Pavelets	1962—1967	" "
25	Tolstovka	1964—1968	" "
26	Khibiny	1963—1968	Coniferous wood
27	Skovorodino	1962—1968	" "
28	Yakutsk	1962—1967	" "
29	Verkhoyansk	1964—1967	" "

To estimate the stability of the data obtained during a relatively small number of observation years, the average monthly number of cases with inversion corresponding to different numbers of observation years were calculated for five stations located in different regional zones. The results are given in Table 2; it is seen that, starting with a 3-year period, the average values are little affected by the length of the averaging period. In all the months of the year they do not, in practice, exceed 1—2 days, i. e., 5—10% (of 147 cases,

TABLE 2. Average number of cases with inversion as a function of the averaging period (1 hour)

Station	Months	Averaging period, number of years				
		3	4	5	6	7
Tamdy	I	22	22	23	23	23
	II	23	22	22	22	22
	III	25	26	26	26	
	IV	26	25	25	25	25
	V	30	29	29	29	29
	VI	29	29	29	29	29
	VII	31	31	30	30	30
	VIII	30	30	30	28	28
	IX	30	30	30	29	29
	X	26	27	28	28	28
	XI	27	27	26	26	26
	XII	23	23	22	22	22
Gigant	IV	22	21	18	19	19
	V	26	26	26	25	24
	VI	24	24	24	24	24
	VII	28	28	28	28	27
	VIII	26	26	25	25	25
	IX	24	24	25	24	24
	X	23	24	24	24	23
Askaniya-Nova	IV	19	19	19	21	21
	V	25	24	24	24	24
	VI	24	23	24	24	24
	VII	28	28	28	28	28
	VIII	28	27	27	27	27
	IX	26	27	27	26	26
	X	21	22	23	23	23
Beregovo	IV	23	23	23	24	24
	V	26	26	26	26	26
	VI	26	26	26	25	26
	VII	27	27	27	27	28
	VIII	27	27	26	26	26
	IX	26	26	25	25	26
	X	23	25	24	25	26
	XI	16	17	17	17	18
Skovorodino	VI	22	23	21	−	−
	VII	22	23	23	24	−
	VIII	24	25	25	25	−
	IX	25	24	22	21	−

in 15 only there is a deviation of 2, and only in 6 cases above 2).
Therefore, even 3-year average samplings are to a certain extent
sufficient to characterize the inversion regime of the ground air
layer.

TABLE 3. Typical annual course of the frequency (%) of temperature inversions of the ground air
layer above a meadow surface in different geographical zones

Zones	I	II	III	IV	V	VI	VII	VIII	IX	X	XI	XII
						Night						
Deserts	60	55	60	75	75	80	80	95	90	70	70	65
Steppes												
ETS*	50	60	60	80	85	85	90	85	85	70	60	50
Kazakhstan	50	60	65	70	80	80	90	85	70	60	60	45
Siberia and Far East .	60	70	80	80	80	90	85	80	85	80	65	40
Woods												
mixed	50	70	70	80	90	90	90	90	85	80	65	50
coniferous	75	65	85	90	80	90	85	85	80	85	75	70
Wide valleys	90	80	80	90	95	95	100	95	95	95	90	85
						Day						
Deserts	20	15	10	1	1	0	0	0	1	1	5	10
Steppes												
ETS*	20	20	10	1	1	0	0	0	1	3	10	20
Kazakhstan	20	10	15	2	1	1	1	0	0	1	10	15
Siberia and Far East .	25	15	15	10	1	1	0	1	3	5	15	20
Woods												
mixed	35	40	30	10	5	3	1	2	5	5	20	35
coniferous	30	40	40	30	5	5	5	5	5	35	50	60
Wide valleys	15	10	5	5	5	0	0	0	0	3	5	10

* [European territory of Soviet Union.]

Table 3 presents the frequency of the inversions over the year
by day and night in different geographical zones according to the
data of observations in a 3-year period. It is called typical, since
it was obtained by averaging (with an accuracy of 5%) the frequencies
determined by a number of stations within the limits of the given
regional zone. It can be seen from the data that at night a ground
temperature inversion, with frequency from 40—50 to 100%, is
observed throughout the year in all the regional zones. Inversions
are observed most frequently in the warm part of the year (their
frequency in all the zones is higher than 70—80%, and falls to 30—40%
in the cold period). The probability of inversions is lowest in the

TABLE 4. Diurnal course of the frequency (%) of temperature inversions of the ground air layer above a meadow surface in different geographical zones

Zones	Month	Observation period (hours)					
		1	7	10	13	16	19
Deserts (Aidarly)	IV	80	25	2	0	1	85
	V	80	10	1	1	5	80
	VI	85	15	0	0	0	65
	VII	85	15	1	0	1	70
	VIII	90	20	1	0	1	80
	IX	90	40	1	1	1	90
	X	80	80	5	2	15	90
Steppes							
ETS* (Askaniya-Nova)	V	85	15	15	5	15	85
	VI	80	10	5	5	5	70
	VII	95	5	1	1	5	65
	VIII	85	10	1	1	3	80
	IX	90	20	2	0	3	90
	X	75	45	5	5	20	85
Kazakhstan (Tselinograd)	V	90	10	5	2	5	65
	VI	80	3	1	0	5	30
	VII	80	2	1	0	1	25
	VIII	75	10	1	1	2	65
	IX	65	20	3	3	5	80
	X	65	50	5	5	35	75
Siberia and Far East (Solyanka)	VI	90	10	5	5	10	60
	VII	90	1	1	1	5	55
	VIII	90	5	1	1	5	90
	IX	90	50	5	5	15	90
Woods							
mixed (Pavelets)	V	90	10	0	1	10	60
	VI	85	5	0	0	5	40
	VII	85	1	0	1	10	60
	VIII	90	15	1	1	10	80
	IX	85	30	5	5	15	90
	X	70	50	5	5	25	70
coniferous (Skovorodino)	VI	75	10	10	0	20	75
	VII	80	10	10	0	35	80
	VIII	90	15	5	0	40	95
	IX	90	40	5	5	40	95
Wide valleys (Dushanbe)	I	90	90	20	15	75	95
	II	100	80	10	5	35	90
	III	90	80	10	5	30	90
	IV	100	50	10	5	50	90
	V	95	30	5	10	50	95
	VI	100	30	5	10	55	95
	VII	100	15	1	2	15	95
	VIII	100	20	1	0	15	100
	IX	95	25	0	1	15	100
	X	95	70	5	1	40	95

winter months (December—January). No appreciable difference
between the frequency of inversions in different regional zones is
noted. However, at some stations located in wide valleys (Dushanbe,
Telavi, Frunze), the frequency of the inversions at night is between
80 and 100% for all months of the year. It is also high in woods,
in particular, coniferous woods. In low humidity zones (deserts,
steppes) the frequency of inversions varies appreciably more than
during the year (the amplitude is 35—40%).

During daytime inversions of the air temperature are observed
in the cold half of the year only; in summer the probability of their
appearance is low. During the cold period, the frequency of inversions varies in different regional zones, and depends on the character of the underlying surface. In the zone of coniferous woods it is
about 5% even in summer, and reaches 30—60% in the period between
October and April, when snow covers the surface. In the zone of
mixed woods the period when the frequency of inversions is 20—40%
is reduced to November—March. The lowest values of the frequency
are observed in the zone of wide mountain valleys, and are 10—15%
in December—February only, while in the other zones such values
are observed in November—March only.

Table 4 shows the diurnal course of the frequency of inversions.
The results were obtained in the summer period (when measurements were taken six times daily), from observations over five
years at individual stations located in the given regional zones.
The table shows that the frequency of inversions varies with the
hour of the day. It is highest at night (1.00 hrs), and also in the
evening (19 hrs). The lowest values are observed during the day
(10, 13, 16 hrs), while in the morning (7 hrs), particularly in autumn,
the frequency reaches 50% or more. The curve of the frequency as
a function of the hour of the day varies considerably from month to
month, and the variation is greatest in spring and autumn, as is
clearly seen from the Dushanbe example.

To characterize the intensity of inversions, temperature
differences greater than 0.5° (i. e., $\Delta t < -0.5°$) were selected in the
0.5—2 m air layer. The frequency of these inversions are given in
Table 5 for night only, since during the day (13 hrs) inversions of
such intensity are observed very rarely, and only during the cold
time of the year. At night, as seen from Table 5, their frequency is
considerable, and varies in different regional zones and at different
times of the year. The highest frequency of inversions with values
higher than 0.5° is observed in all regional zones in summer; during
the cold period of the year it is lower by a factor of 2—3 than during
the warm period. In drier regions (deserts, steppes) the frequency
is lower than in more humid ones (coniferous wood, wide valleys).
But the amplitude of the yearly fluctuations is greater in the dry regions.

It was found that the air temperature does not affect the formation of inversions: the same values for the probability of inversions are noted for any values of the air temperature, and the differences are determined only by this or that temperature limit in the given time period.

TABLE 5. Frequency of night inversions for $\Delta t_{0.5-2} <- \ 0.5°$ (from data of a three-year period)

Zones	I	II	III	IV	V	VI	VII	VIII	IX	X	XI	XII
Deserts	17	10	10	15	19	23	25	30	26	18	17	19
Steppes												
ETS	6	8	10	18	30	35	36	27	27	19	6	2
Kazakhstan	0	6	8	15	14	26	20	16	20	6	8	3
Siberia and Far East	17	24	26	28	26	26	25	18	24	18	10	10
Woods												
mixed	9	9	10	21	28	35	34	31	22	15	8	6
coniferous	30	17	43	47	34	41	36	34	31	36	25	18
Wide valleys	56	56	41	46	52	65	70	61	57	47	61	55

TABLE 6. Frequency of night inversions for different wind speeds u_2

Zones	u_2 (m/sec)	I	II	III	IV	V	VI	VII	VIII	IX	X	XI	XII
Deserts	<2	20	25	20	30	30	30	35	35	40	30	30	20
	>4	10	15	15	15	15	15	15	20	15	10	10	15
Steppes													
ETS	<2	15	25	25	35	55	60	65	55	60	40	20	15
	>4	10	10	10	10	5	1	1	2	5	2	5	10
Kazakhstan	<2	15	25	30	40	35	45	45	45	40	25	35	20
	>4	10	15	15	10	15	10	10	5	10	15	10	5
Siberia and Far East ..	<2	35	45	45	55	50	65	75	65	65	55	40	20
	>4	5	5	5	5	5	1	1	1	1	2	5	5
Woods													
mixed	<2	25	30	35	40	50	70	65	70	55	40	30	20
	>4	10	10	10	5	5	1	1	1	2	5	5	10
coniferous	<2	45	40	65	60	65	70	70	65	65	70	55	45
	>4	5	10	5	5	1	1	0	0	1	5	5	5
Wide valleys	<2	75	80	65	80	80	85	90	85	90	85	75	60
	>4	2	2	1	0	0	1	1	1	0	3	1	1

The wind speed has a greater effect on the formation of inversions, as seen in Table 6, which shows the frequency of inversions at night for different months of the year at wind speeds below 2 m/sec and above 4 m/sec. At night, in the warm part of the year over a considerable part of the USSR, more than 50% of the inversions occur at wind speeds below 2 m/sec, and only in the desert zone and the steppes of Kazakhstan the frequency of inversions for this speed falls to 30—45%, i. e., in these regions inversions occur mainly at speeds above 2 m/sec. At wind speeds above 4 m/sec, the frequency of inversions does not exceed 10—15%, mainly in the desert and steppe regions. During the cold period of the year the frequency of inversions increases at wind speeds above 2 m/sec.

TABLE 7. Frequency of inversions in cloudy weather

Zone	Hour of the day	I	II	III	IV	V	VI	VII	VIII	IX	X	XI	XII
Mixed wood .	10	—	—	—	—	4	4	8	6	10	10	6	—
	13	12	12	5	2	3	2	8	7	12	10	11	20
	16	—	—	—	—	30	12	14	14	17	29	42	—
Steppe	10	—	—,	—	6	10	8	4	7	8	8	—	—
	13	11	8	16	0	6	6	9	5	8	8	13	12
	16	—	—	—	—	18	10	13	10	20	24	—	—

TABLE 8. Mean values of Δt_{0-2} for different values of $\Delta t_{0.5-2}$ and a wind speed $u_2 < 2$ m/sec

Station	$\Delta t_{0.5-2}$	Month			
		I	IV	VII	X
Beki-Bent	0.0, -0.2	-0.5	-0.7	-0.6	-2.1
	-0.3, -0.5		-1.4	-0.8	-2.5
	-0.6, -1.0		-2.2	-2.9	-2.8
	<-1.0	-5.7			-4.0
Askaniya-Nova	0.0, -0.2		-0.5	-0.9	-2.5
	-0.1, -0.5		-1.3	-1.3	-2.1
	-0.6, -1.0		-1.6	-2.2	-3.5
	<-1.0			-3.2	-1.4
Solyanka	-0.0, -0.2			0.0	-0.4
	-0.3, -0.5	-4.6	-1.9	0.0	-1.4
	-0.6, -1.0	-7.3	-4.6	-1.3	-3.4
	<-1.0		-5.7	-2.3	-6.4
Nikolaevskoe	0.0, -0.2		-2.2	0.2	-0.2
	-0.3, -0.5		-3.3	-0.3	-0.7
	-0.6, -1.0		-3.5	-0.4	
	<-1.0		-3.6	-0.7	
Skovorodino	0.0, -0.2		-1.4	-0.2	-0.7
	-0.3, -0.5	-1.6	-2.8	0.4	-0.4
	-0.6, -1.0		-3.1	0.2	-0.9
	<-1.0		-5.1		-0.2

The results of the observations indicate that temperature inversions in the ground air layer can also occur in cloudy weather. Table 7 gives the frequencies of inversions in 30—70 cases of observations in cloudy weather over the period 1962—1968. It is seen that in cloudy weather the frequency of inversions is low, usually less than 10%, but it increases in summer in the afternoon, and also during the cold period of the year.

To clarify the connection between the temperature inversions in the 0—2 and 0.5—2 m layers, the corresponding values of the temperature differences observed during 1965—1967 at five stations in different regional zones were studied. The results are presented in Table 8 as mean values of the soil-air temperature difference (Δt_{0-2}) for different limits of the air temperature gradients ($\Delta t_{0.5-2}$) at wind speeds below 2 m/sec. Only mean values obtained from not less than five cases of separate observations are included. Table 8 shows that temperature inversions occur only if there is a temperature drop in the air of between -0.5° and -2.5° even for weak inversions (up to -0.2°). The temperature drop increases with increase in the intensity of inversions, and increases by a factor of 2—3 for the inversion limits which we selected between 0.0—0.2 and -1°. The value of the temperature drop corresponding to a given intensity of inversion varies during the year: it is higher in winter (when there is snow cover); the annual change is smaller when there is no snow cover (the Beki-Bent station).

The results given in this paper give a satisfactory description of the behavior of the air temperature inversions in the two-meter thick layer.

Bibliography

1. Rukovodstvo po gradientnym nablyudeniyam i opredeleniyu sostavlyayushchikh teplovogo balansa (Handbook for Gradient Observations and the Determination of the Heat-Balance Components). Leningrad, Gidrometeoizdat. 1964.
2. Vasil'eva, L.G. and T.A.Golubova. Prizemnaya inversiya temperatury vozdukha v raznykh raionakh SSSR (Ground Inversion of Air Temperature in Different Regions of the USSR). — Trudy GGO, No. 264, pp. 3—22. 1970.

VERTICAL STRUCTURE OF THE TURBULENCE ENERGY ABOVE A RUFFLED SURFACE

E. L. Genikhovich, L. Yu. Preobrazhenskii

From experimental studies of the wind structure above the sea carried out recently, the turbulence characteristics in the layer above the water can be determined. The results indicate considerable differences between the structure of the turbulent stream above the sea and its structure in the ground layer of the atmosphere. In particular, in indifferent stratification, the ratio of the turbulence energy to the square of the dynamic velocity is not a universal constant /3, 4, 8/. However, from the measurements we cannot describe in sufficient detail the vertical distribution of the turbulence energy, which must depend on the characteristics of the roughness of the sea. This is confirmed by the particular form of energy spectra, and by the form of the correlation functions of the fluctuations in wind speed /1, 3, 8/.

The fluctuations in the longitudinal u' and vertical w' components of the wind speed were measured on the sea expeditions of GGO [Main Geophysical Observatory] in the North Atlantic with a Froude-type spar buoy, and in the Baltic sea with a land-based structure. The measurements were carried out by means of thermoanemometers placed at levels of 1, 2, and 5 m. The results show that the variances in the longitudinal and vertical wind components, σ_u^2 and σ_w^2, depend on the sea roughness and on the height above the mean sea level.

As an example, in Figure 1 we plotted the vertical profiles of σ_u and σ_w as a function of the dimensionless height z/λ, where λ is the mean wavelength of 50% probability, for values of the dynamic velocity v_* 0–9, 10–19, and 20–40 cm/sec. (This corresponds approximately to aerodynamically smooth, intermediate, and rough modes of flow past the water surface /5/.)

The results given, and also the results of spectra measurements /3/, indicate that the flow past a ruffled surface leads to an additional generation of turbulent energy, due to the bending of the streamlines and the change in the velocity profile curvature near the boundary. It seems advisable to build a simple mathematical model of this process.

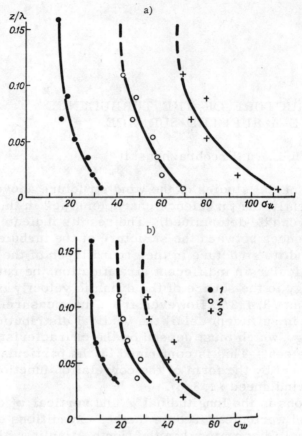

FIGURE 1. Vertical profiles of the mean square fluctuations
of the longitudinal (a) and vertical (b) components of the
wind speed above the sea for different values of v_*:

1) $v_* = 0-9$ cm/sec; 2) $v_* = 10-19$ cm/sec; 3) $v_* = 20-40$ cm/sec.

    To describe the behavior of the air stream near a bent underlying
surface, in the equations of motion we must convert to flow coordinates (the stream function and the velocity potential of the
potential flow in the given region). The equations are simplified
by the standard methods of the boundary layer theory /2/; with a
small slope of the waves it can be shown that to a first approximation the wind speed at some level above the surface depends only
on the value of the stream function of the potential flow at that level.
In this case, for the velocity component along the streamline we
use the simplest model

$$u = \frac{v_*}{\varkappa} \ln \frac{\psi}{\psi_0},$$

(1)

where ψ is the value of the stream function, ψ_0 the roughness of flow of the underlying surface, $\varkappa = 0.4$, the von Karman constant.

The character of the vertical distribution of the time-averaged turbulence energy can be determined from the balance equations of the turbulence energy.

If we assume that the ruffled surface is homogeneous in the direction perpendicular to the wind, and that the coefficient of eddy exchange is isotropic the transformation component Tr of the turbulence energy balance can be written in the form

$$\mathrm{Tr} = k \left\{ \left(\frac{\partial u}{\partial z} \right)^2 + 4 \left(\frac{\partial u}{\partial x} \right)^2 \right\}. \tag{2}$$

It is known (cf. /8/) that the stream function for a potential flow above a flowing sinusoidal surface has the form

$$\psi = u_\infty \left\{ z + a e^{-\frac{2\pi z}{\lambda}} \cos\left(\frac{2\pi x}{\lambda} - \omega t \right) \right\}, \tag{3}$$

where u_∞ is the velocity of the unperturbed impinging flow, a the amplitude, λ the wavelength, ω the cyclic frequency of the flowing wave. By substituting (1) and (3) in (2), and averaging the expression obtained with reference to time, we obtain

$$\mathrm{Tr} = k u_\infty^2 \left(\frac{du}{d\psi} \right)^2 \left(1 + \frac{10\,\pi^2 a^2}{\lambda^2} e^{-\frac{4\pi z}{\lambda}} \right). \tag{4}$$

By representing the coefficient of eddy exchange in a form conforming with (1)

$$k = \varkappa v_* \frac{\psi}{u_\infty}, \tag{5}$$

we obtain

$$k u_\infty \left(\frac{du}{d\psi} \right) = v_*^2. \tag{6}$$

If we neglect the diffusion term, we can write the balance equation of the turbulence energy in the form

$$\mathrm{Tr} = \mathrm{Di}, \tag{7}$$

where the dissipation term Di is determined from Kolmogorov's formula

$$\mathrm{Di} = \frac{c b^2}{k}, \tag{8}$$

in which $c = 0.046$, according to /7/. The following final expression for the turbulence energy is obtained from (4), (6), and (8):

$$\frac{b}{v_*^2} = \frac{1}{Vc}\left(1 + 10\frac{\pi^2 a^2}{\lambda^2}\, e^{-\frac{4\pi z}{\lambda}}\right)^{\frac{1}{2}}. \tag{9}$$

It follows from (9) that the turbulence energy b is proportional to the dynamic velocity v_* and decreases with the distance from the boundary of separation.

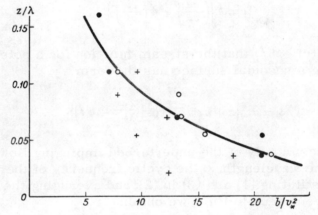

FIGURE 2. Vertical distribution of the turbulence energy in the layer adjacent to the water (curve calculated from formula (10)). Legend as in Figure 1.

However, the numerical factors entering this expression are determined from a crude model of the flow, which does not allow for factors such as the nonsinusoidal shape of the ruffled surface, the transformation of the velocity profile along the individual wave, and the resulting possible variation in the dynamic velocity v_*. We assume, therefore, that the variation in the turbulence energy with height is described by a relation of type (9):

$$\frac{b}{v_*^2} = \frac{1}{Vc}\left(1 + A\frac{a^2}{\lambda^2}\, e^{-B\frac{z}{\lambda}}\right)^{\frac{1}{2}}, \tag{10}$$

where the coefficients are determined from an empirical graph (Figure 2) representing b/v_*^2 as a function of the dimensionless height z/λ, in which b was determined from the formula

$$b = \frac{2\sigma_u^2 + \sigma_w^2}{2}.$$

A curve calculated for empirical values of the constant $A = 1 \cdot 10^5$, $B = 32$, is given in the same figure, with the ratio a/λ taken as $0.25 \cdot 10^{-1}$, which corresponds to the slope of large ocean waves. A comparison of formulas (9) and (10) shows that for real flows the rate of attenuation of the boundary perturbation with height is apparently greater than the value predicted from the theory of potential flows. Thus, model (1) for the wind speed can be considered satisfactory if in formula (3) we replace the coefficient 2π in the exponent by a higher value.

Bibliography

1. Byshev, V. I. and O. A. Kuznetsov. Strukturnye kharakteristiki atmosfernoi turbulentnosti v privodnom sloe nad
   otkrytym okeanom (Structural Characteristics of Atmospheric Turbulence in the Bottom Layer above the
   Open Sea). — Izvestiya AN SSSR, Fizika Atmosfery i
   Okeana, Vol. 5, No. 6. 1969.
2. Van Dyke, M. Perturbation Methods in Fluid Mechanics
   New York-London, Academic Press. 1964.
3. Volkov, Yu. S. Spektry pul'satsii skorosti i temperatury
   vozdushnogo potoka nad vzvolnovannoi poverkhnost'yu
   morya (Spectra of the Fluctuations in Velocity and
   Temperature of the Air Flow above a Rough Sea Surface). —
   Izvestiya AN SSSR, Fizika Atmosfery i Okeana, Vol. 5,
   No. 12. 1969.
4. Zubkovskii, S. L. and T. K. Kravchenko. Pryamye izmereniya nekotorykh kharakteristik atmosfernoi turbulentnosti
   v privodnom sloe (Direct Measurements of Some Characteristics of Atmospheric Turbulence in the Air Layer above
   the Water). — Izvestiya AN SSSR, Fizika Atmosfery i
   Okeana, Vol. 3, No. 2. 1967.
5. Kitaigorodskii, S. A. Melkomasshtabnoe vzaimodeistvie
   okeana i atmosfery (Small-Scale Interaction between the
   Ocean and the Atmosphere). — Izvestiya AN SSSR, Fizika
   Atmosfery i Okeana, Vol. 5, No. 11. 1969.
6. Milne-Thomson, L. M. Theoretical Hydrodynamics. London,
   Macmillan. 1960.
7. Monin, A. S. and A. M. Yaglom. Statisticheskaya gidromekhanika. Ch. 1. (Statistical Hydromechanics. Part 1).
   Moskva, "Nauka." 1967.

SOME POSSIBILITIES OF FORECASTING THE CONCENTRATION OF CONTAMINANTS IN CITY AIR

L. R. Son'kin

The concentration of contaminants in the air is highest when large amounts are discharged under meteorological conditions unfavorable for their dispersion. It is therefore necessary to be able to forecast such conditions so as to avoid excessive discharges in such cases.

The development of a forecasting method is related to the study of the meteorological factors causing a considerable concentration of contaminants in the ground air layer.

Meteorological conditions under which an increased concentration of contaminants is observed are known for some particular cases. Thus, in towns in which the industrial areas lie beyond the residential quarters, a necessary condition for an increased concentration of contaminants is a suitable wind direction. The meteorological factors leading to the highest concentration of contaminants with this wind direction have been analyzed in theoretical studies /2—5/. These factors are related to the dangerous wind velocity, depending on the discharge parameters, raised inversion located above the sources, fogs. The conclusions derived from the theoretical relationships have been implemented in some towns (Dzerzhinsk, Zaporozh'ye).

It is much more difficult to establish the meteorological conditions of higher contaminant concentration in towns which have a large number of different discharge sources within their limits. It is practically impossible to take into account the discharge parameters of all the sources, especially since many of them are random. The problem becomes statistical. The correlation between the existence of contaminants in city air and the meteorological conditions, and also the synoptic processes, has been analyzed in a number of papers /11—16, et al./, including earlier studies by the author /6—9/. Particular statistically significant relations were obtained, which mainly hold, but there are appreciable exceptions in specific cases. When these relations were later made more accurate, attention was paid mainly to a study of the meteorological

conditions of the total content of contaminants in the air above the
whole town.   It was characterized by means of the parameter

$$P = \frac{m}{n},$$

where n is the total number of observations during the time interval
studied (one or several days), m the number of observations during
the same period in which the concentrations were 1.5 times the
average seasonal values for each station separately /10/.   Obviously, the parameter P can vary from 0 to 1.   On an average, $P \approx 0.2$.
     The meteorological conditions for a content of contaminants in
the air over the town as a whole are studied separately for the
different seasons of the year.   We give here the results of the
analysis of data for the cold half-year of 1967—1968 in Leningrad,
Sverdlovsk, Alma-Ata, Kursk, Krasnoyarsk, and also for two cold
half-years (1967—1968 and 1968—1969) in Chita.   In a study of the
content of contaminant in the air under stagnant conditions, the data
of Irkutsk and Novisibirsk for the period October 1967—March 1968
were also used.
     The introduction of the parameter P is useful for making the
relations more accurate.   In fact, the information on the concentration of contaminants in the air contains a random component,
due to the nonperiodic fluctuations of the discharges and some
other factors, and this can lead to a deviation from the established
relationships.   The parameter P is less sensitive to random fluctuations of the discharges than the values of the individual concentrations /10/, and is more sensitive to the meteorological
conditions.   Therefore, when we use P we eliminate at least one of
the factors leading to deviations from the relationships.   However,
it was found that in spite of this the relations between the meteorological conditions and the contaminant content in the air found earlier
are not always implemented in specific cases.   Thus, it is known
that one of the dangerous conditions in towns is air stagnation,
characterized by very weak winds (0—1 m/sec) and ground inversions.
An analysis of the materials from the eight towns listed showed that
the frequency of high contaminant content in the air ($P > 0.2$) under
stagnant conditions is only 70%.   At the same time large values of P
were frequently found under weather conditions which are not
always considered to be unfavorable for large towns.
     It follows that the deviations from the statistical relations are
due not only to the existence of a random component in the information, but also to neglect of some essential meteorological factors.
In a later analysis, the effect of the known relations is studied in
greater detail, and an attempt is made to clarify and consider new factors.

Firstly, it is of interest to find the average variation in the
content of contaminants in the air when the stagnant conditions
(a combination of calm weather and ground inversion) are disturbed
by an increase in the wind speed and the elimination of inversion,
and to note which of the factors is more important.   It is shown
that an increase in wind speed to 1—2 m/sec reduces the content of
contaminants in the atmosphere less than the elimination of the
ground inversion in calm weather (Table 1).   The table does not
include the results for Kursk, since there were not many cases with
no wind.

TABLE 1. Index of content of contaminants in the air under near-stagnant conditions

Town	Wind speed (m/sec) in the presence of ground inversion			Calm weather and absence of ground inversion
	calm weather	1	2	
Leningrad	0.27	0.23	0.19	0.14
Sverdlovsk	0.18	0.17	0.18	0.12
Alma-Ata	0.37	0.33	—	0.17
Krasnoyarsk	0.33	0.33	0.30	0.19
Chita	0.25	0.23	0.16	0.20

The results given in Table 1 agree with the conclusions of
Berlyand's theoretical investigations /3/, from which it follows
that a weakening of the wind to calm has a double effect on the
content of contaminants in the air.   On the one hand, the concentration of contaminants in the ground layer of the atmosphere must
increase greatly.   At the same time, the ascent of heated discharges increases without restriction.   In the absence of wind the
discharges rise to higher layers of the atmosphere and are dispersed.   However, if the calm is accompanied by inversion, a ceiling
which prevents further ascent of the rising discharges is created.
As a result, the concentration of contaminant in the ground layer
increases greatly.   This is the reason why, as seen from Table 1,
in calm weather the content of contaminants in the air drops sharply
in the absence of inversions: if the calm is not accompanied by a
high stability of the atmosphere, even slightly overheated discharges are transferred to higher layers and do not contaminate
the ground air layer.
Thus, the presence of calm or of a weak wind cannot be considered by itself as an unfavorable situation.   The frequency of

weak winds can justifiably be considered as a characteristic for evaluating the potential of air contamination in winter over a large territory /1/ because these winds are usually accompanied by stable stratification.

The wind speed and the stability of the atmosphere do not determine fully the character of contaminant content in the air, and an allowance for these parameters is therefore not sufficient for its forecast. In analyzing the other factors, it seemed natural to consider firstly the wind direction. When the observations of contaminant concentration at different points of a number of towns were processed, it was shown that in most cases there is no connection between the contaminant content and the wind direction /6/. This is due to the characteristics of the distribution of contaminants in a town: a large number of sources with different discharge characteristics, a complex wind field, the creation of a background concentration. In our study, when we processed the values of P characterizing the contaminant content in the air over the town as a whole, we showed that with certain wind directions the average content of contaminants in the city atmosphere increases. This is seen in Table 2.

TABLE 2. Frequency of cases with a higher content of contaminants in the air ($P > 0.2$) for some wind directions

Town	Unfavorable wind directions	Frequency (%)
Leningrad	E, SE, S, SW	79
Sverdlovsk	SE, S	68
Kursk	N, NE, E	58
Krasnoyarsk	NE, E	83
Alma-Ata	SE, E	80
Chita	N, NE, E, SE	60

Thus, the wind direction can be considered as one of the independent factors determining air pollution in a town.

The accumulation of contaminants in the town air depends on the air temperature in the ground layer: other conditions being equal, a higher contamination is observed at a relatively high temperature. This is frequently difficult to detect, due to the superposition of other factors. Thus, a considerable accumulation of contaminants is caused by air stagnation, which is usually observed at very low temperature. Very active winter cyclones cause a considerable increase in the temperature, and at the same time clear the atmosphere of contaminants. Therefore, if we consider the connection

between the concentration of contaminants in the air and its temperature independently of the other weather conditions, we note
frequently a decrease instead of an increase in contaminant content
with increase in temperature, caused by the phenomena listed above.

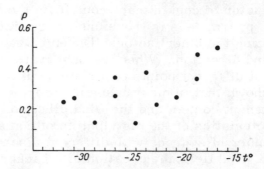

FIGURE 1. Relationship between content of contaminant in the air (parameter P) and the diurnal
temperature under conditions of high stagnation.
Chita.

Thus, we must consider the connection between the pollution of
the air and its temperature under roughly the same weather conditions.   This was possible for Alma-Ata and Chita, where weak
winds dominate, and variation in the meteorological conditions is
relatively low.   From the results for the three winter months at
Alma-Ata (December 1967, January and February 1968) 18 cases
were selected in which the ground inversion was accompanied by a
wind speed of 0−1 m/sec.   The relationship between P and the
maximum air temperature of the day (t_M) was studied for these
cases.   It was found that there exists a definite positive correlation,
expressed by a correlation coefficient $r = 0.47 \pm 0.19$.   For $t_M > 0$,
$P = 0.43$ on an average, for $t_M < 0$, $P = 0.24$.
    From the observations in Chita 12 cases were selected in which
under conditions of air stagnation a strong ground inversion was
maintained during the day (thickness at least 500 m and intensity at
least 5°).   The correlation with the air temperature is even higher
here (Figure 1), $r = 0.66 \pm 0.17$.
    The trend toward a higher content of contaminants in the air at
relatively high temperatures is noted not only under stagnant conditions.   Thus, in Leningrad during winter thaws and weak winds the
content of contaminants in the air was higher ($P > 0.2$) in 7 cases out
of 8; at Sverdlovsk at an air temperature in winter above -10° and a
wind speed of 1−2 m/sec, in 10 cases out of 13; at Krasnoyarsk at an
air temperature above 8°, and a wind speed at not more than 5 m/sec,
in 9 cases out of 11.

There are two possible reasons for the direct correlation between contaminants in city air and its temperature. In the first place, under stagnant air conditions at low winter temperatures, the effect of an island of warmth is heightened, leading to an increased influx of relatively pure air as a result of local circulation. Zil'bershtein /5/ also came to a similar conclusion. A second reason may be an increase in the buoyancy of overheated discharges with a drop in temperature as a result of increase in the temperature difference between the discharged contaminants and the surrounding air.

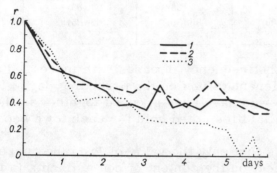

FIGURE 2. Autocorrelation time function of parameter P:

1) Krasnoyarsk; 2) Alma-Ata; 3) Chita.

The time-lag factor has a considerable effect on contaminant content in town air. The contaminant content in town air as a whole is very dependent on its value on the preceding days, and this is frequently the reason for disturbance of the known relationships.

To study the time-lag factor, we analyzed the time correlation function of the magnitudes P, calculated on a Ural-4 computer according to a specially written program. The shape of the function for a number of towns is shown in Figure 2. Since the observations on the content of contaminants were carried according to a sliding curve (Monday, Wednesday and Friday at 15, 18, and 21 hours; Tuesday, Thursday and Saturday at 6, 9, and 12 hours), the time invervals between the mean observation periods were (in days): $2/3$, $1\frac{1}{3}$, 2, $2\frac{1}{3}$, $2\frac{2}{3}$, 3, $3\frac{1}{3}$, $3\frac{2}{3}$, etc. Figure 2 shows that a stable positive correlation is maintained in the course of 4—5 days. For an interval of $2/3$ day $r = 0.7-0.8$, for an interval of 4—5 days, $r = 0.25-0.30$.

We shall consider the time-lag factor in a form convenient for forecasting.

TABLE 3. Frequency (%) of higher content of contaminants in the air ($P > 0.2$) as a function of the value of P on the preceding day (the number of cases of corresponding sampling is given in parentheses)

P_{n-1}	Observations carried out			
	on current day in morning, on preceding day in evening		on current day in evening, on preceding day in morning	
	in general case	during air stagnation	in general case	during air stagnation
< 0.1	11 (105)	24 (33)	24 (66)	50 (10)
≥ 0.3	92 (109)	95 (39)	66 (75)	100 (14)

Table 3 gives the frequency of cases with increased air contamination after different values on the preceding day. The results for the different towns have been grouped together in the table. However, the peculiarities noted hold for each town separately as well.

The table shows that only the value of P on the current day has a considerable effect on the content of contaminants in the town air on the following day. A very close connection is observed between the contaminant content in the air on the current day from the results of morning observations, and on the preceding day from the results of evening observations. Table 3 also gives values of the contaminant content in the air under stagnant conditions as a function of the value of P on the preceding day (P_{n-1}). An interesting fact is that after a day with relatively pure air $(P_{n-1} < 0.1)$ and under stagnant conditions no increased contaminant content is observed. Most of the cases when the air was relatively pure after stagnation fall into this category. Thus, two types of time-lag are observed; meteorological, and of the content of contaminants in the air itself. Meteorological time-lag implies a tendency to preserve the atmospheric factors which determine the level of the content of contaminants in the air. A very important fact is that some of these may may be unknown, and are allowed for automatically to a certain extent by considering the previous content of contaminants in the air. The relatively pure air under stagnant conditions after a day with low values of P is determined by the time-lag of the content of contaminants in the air itself.

During air stagnation periods, the effect of the time-lag factor in the form described above is disturbed. No accumulation of contaminants in the town air takes place in these periods. From the results of a number of towns we found 21 cases of air stagnation in winter lasting not less than three days. A continuous increase in the content of contaminants in the air was observed in two cases only, and in both cases the stagnation lasted only three days. Table 4 gives an idea of the variation in the content of contaminants in the air during the stagnant periods studied.

TABLE 4. Number of cases with different variations in the content of contaminants in the air during long stagnant periods (3 days or more)

Total number	Continuous growth	Continuous fall	One maximum in the middle period	2—3 maxima during the period	Constant
21	2	7	4	6	2

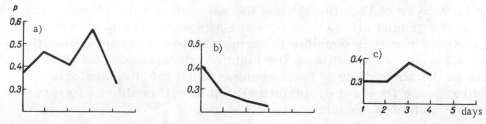

FIGURE 3. Characteristic types of variation in the content of contaminants in the air during an air stagnation period:

a) Krasnoyarsk, 30.1.68—3.2.68; Chita, 25.12.67—28.12.67; c) Krasnoyarsk, 6.12.67—9.12.67.

Thus, during the air stagnation period we usually observe either a fluctuation in the content of contaminants in the atmosphere with one or several maxima, or a decrease. Characteristic examples of variation in contaminant content in the air during a stagnant period are shown in Figure 3. Obviously, under given conditions there are factors preventing the continuous growth in the content of contaminants in the air. Firstly, a dynamic equilibrium can be established at any moment of time between the discharged and dispersed contaminants, since the increase in contaminant concentration in the town is accompanied by the dispersion of the contaminants due to an increase in the concentration gradient. Secondly, under stagnant conditions, the effect of an island of warmth in the town is heightened as a result of both accumulation of contaminants and drop in temperature, and this leads to increased turbulent exchange. It was found that the most dangerous cases of the appearance of stagnation are those when on the day preceding the stagnation a high value of P was observed. From the data of the two winter seasons in Chita

19 such cases $(P_{n-1}>0.25)$ were selected with $P=0.34$ on an average. If a stagnant condition is not observed the first day, then when $P_{n-1}>0.25$, according to the data of Chita the average content of contaminants in the air is not nearly as high, $P=0.28$.

Thus, the moment of drop in the wind speed to values below 1 m/sec and the formation of stagnation conditions is dangerous. Under these conditions, 12 cases (out of 44) were observed with highest contents of contaminants in the air $(P>0.4)$. The data given show that there are a number of meteorological factors determining the concentration of contaminants in city air. These factors interact with one another, and with the factors of propagation of the discharges in town. However, it is possible to isolate several meteorological factors which interact simultaneously, and at the same time (to a certain extent) independently, and determine contents of contaminants in the town air: the wind speed, the stability of the boundary layer of the atmosphere, the wind direction, the temperature of the ground air layer before the air contamination. From the results obtained it is possible to formulate, to a first approximation, the meteorological conditions for higher concentrations of contaminants in the air for use in forecasting. Since the meteorological conditions are forecast systematically, this will enable us to forecast the content of contaminants in the air.

Rules for forecasting higher contents of contaminants in the air, and rules for forecasting the most dangerous accumulation of contaminants in town air, have been formulated. In the first case, the aim is an alternative forecast of a higher content of contaminants in the air, defined arbitrarily as $P>0.2$. Obviously, while such forecasts are of a certain interest, their practical importance is limited. The most interesting are the forecasts of days with the highest content of contaminants in the air, when it is advisable to decrease the discharges and prevent health hazards.

The following meteorological conditions were established as those determining higher contents of contaminants in the air $(P>0.2)$.

1. Air stagnation (ground inversion and wind speed 0–1 m/sec).

The mean overall frequency of cases with $P>0.2$ is about 70%, as already mentioned. The highest percentage was noted in Alma-Ata (80%), Krasnoyarsk (78%), Irk (79%). The rule formulated above was not confirmed in Sverdlovsk (33%) and Novosibirsk (50%), apparently because of the large number of high discharges, for which stagnation is not a dangerous condition.

2. Air stagnation combined with a relatively high air contamination on the preceding day, $P_{n-1}>0.15$.

The mean overall frequency of cases with $P>0.2$ is 88%. This rule is also confirmed for each town separately.

3. Unfavorable wind directions.

For each of the given towns there are one or several wind directions corresponding to a relatively high content of contaminants. Under such unfavorable wind directions the frequency over all the towns together is 75%.

4. Above-average air temperature during weak winds.

This condition is found in Leningrad, Sverdlovsk, Krasnoyarsk, and Alma-Ata, but the criteria for the different towns proved to be different. The mean frequency of $P > 0.2$ for these four towns was 80%.

5. High level of contaminants in the air on the preceding day, and when sharp weather variations leading to clearing of the atmosphere (a cyclone, a considerable increase in the pressure gradient, etc.) are not expected.

If on the preceding day $P_{n-1} > 0.3$, on the next day the mean frequency of $P > 0.2$ for all the towns together is 80%, independently of the other conditions. The necessity of allowing for possible sharp weather variations when using the time-lag factor is obvious.

There are two types of meteorological conditions of the most dangerous accumulation of contaminants in the town air.

1. Cases of dangerous content of contaminants in the air when several of the above unfavorable meteorological conditions are combined, including $P_{n-1} > 0.25-0.30$.

We give below the conditions for the highest contamination of the air for the different towns.

Krasnoyarsk. Air stagnation or relatively high temperature ($t > 8$) with weak winds combined with $P_{n-1} \geqslant 0.3$. During the winter, 14 days with these conditions were observed, and they included all the cases of highest content of contaminants in the air.

Alma-Ata. Air stagnation at positive maximum temperature combined with $P_{n-1} \geqslant 0.30$. The 8 days with these conditions included all the cases with the most dangerous accumulation of contaminants in the air.

Leningrad. Weak wind (up to 2 m/sec) from unfavorable directions (E, SE, SW) combined with ground inversion. The 16 days with such conditions include all the dangerous cases. If we add the condition $P_{n-1} \geqslant 0.3$, there remain only 6 days with very high content of contaminants in the air.

Sverdlovsk. One of three unfavorable conditions: air stagnation, relatively high temperature ($t > -10$) with a weak wind from an unfavorable direction (SE, S) combined with $P_{n-1} > 0.25$. The 12 days with these conditions include all the cases of dangerous content of contaminants in the air.

Chita.   Ground inversion and calm or very weak wind $(0-1\,\mathrm{m/sec})$ from N, NE, E, or SE, excluding the coldest days $(t<-30°)$ for $P_{n-1}>0.15$.   During the two winter seasons 22 days with these conditions were observed.   These include all the cases of the highest content of contaminants in the air.

2. Dangerous accumulations of contaminants in the air in all the given towns when the wind speed drops below $1\,\mathrm{m/sec}$, and stagnant air with the condition that on the preceding day $P_{n-1}>0.25$.

From these rules it is possible for each of these towns during winter to give about 10 warnings of dangerous concentrations of contaminants in the air.   The municipalities can thus take measures to reduce discharges into the air.

Bibliography

1. Bezuglaya, E. Yu.   K opredeleniyu potentsiala zagryazneniya
      vozdukha (Determination of the Potential of Air Con-
      tamination). — Trudy GGO, No. 234, pp. 69—79. 1968.
2. Berlyand, M. E. et al.   O zagryaznenii atmosfery promysh-
      lennosti vybrosami (Contamination of the Atmosphere by
      Industrial Discharges). — Meteorologiya i Gidrologiya,
      No. 8. 1963.
3. Berlyand, M. E.   Ob opasnykh usloviyakh zagryazneniya
      atmosfery promyshlennymi vybrosami (Dangerous Con-
      ditions of Contamination of the Atmosphere by Industrial
      Discharges). — Trudy GGO, No. 185. 1966.
4. Berlyand, M. E., R. I. Onikul, and G. V. Ryabova.   K teorii
      diffuzii v usloviyakh tumana (Theory of Diffusion under
      Fog Conditions). — Trudy GGO, No. 207. 1968.
5. Zil'bershtein, I. A.   O zagryaznennosti atmosfernogo
      vozdukha v Chite (Pollution of the Atmosphere in Chita). —
      Okhrana prirody i vosproizvodstvo estestvennykh resur-
      sov, No. 1. Chita. 1967.
6. Son'kin, L. R., E. A. Razbegaeva, and K. N. Terekhova.
      K voprosu o meteorologicheskoi obuslovlennosti zagryaz-
      neniya vozdukha nad gorodami (Problem of the Meteoro-
      logical Factors Determining City Air Pollution). — Trudy
      GGO, No. 185. 1966.
7. Son'kin, L. R.   Nekotorye rezul'taty sinoptiko-klimatolo-
      gicheskogo analiza zagryazneniya vozdukha v gorodakh
      (Some Results of the Synoptic-Climatological Analysis of
      City Air Contamination). — Trudy GGO, No. 207. 1968.

8. Son'kin, L.R. and D.V. Chalikov. Ob obrabotke i analize
   nablyudenii za zagryazneniem vozdukha v gorodakh (Processing and Analysis of Observations of City Air Contamination). — Trudy GGO, No. 207. 1968.

9. Son'kin, L.R.   Analiz meteorologicheskikh uslovii opasnogo
   zagryazneniya vozdukha v gorodakh (Analysis of the
   Meteorological Conditions of Dangerous Contamination of
   City Air). — Trudy GGO, No. 234. 1968.

10. Son'kin, L.R. and T.P. Denisova. Meteorologicheskie
    usloviya formirovaniya periodov intensivnogo zagryaz-
    neniya vozdukha v gorodakh (Meteorological Conditions of
    the Formation of Periods of High Contamination of City
    Air). — Trudy GGO, No. 238, pp. 33—41. 1969.

11. Shevchuk, I.A., L.I. Vvedenskaya, and L.I. Volodkevich.
    Povtoryaemost' meteorologicheskikh uslovii, sposobst-
    vuyushchikh uvelicheniyu zagryazneniya prizemnogo sloya
    atmosfery (Frequency of the Meteorological Conditions
    Leading to Increased Pollution of the Ground Layer of the
    Atmosphere). — Trudy Novosibirskogo Regional'nogo
    GMTs, No. 2. 1969.

12. Dickon, R.R.   Meteorological Factors Affecting Particulate
    Air Pollution of a City. — BAMS, Vol. 42, No. 8. 1961.

13. Harada, H.   Relation between Air Pollution in Osaka and
    Anticyclone. — J. Met. Res., Tokyo, Vol. 20. No. 12. 1968.

14. Holzworth, G.C.   Large-Scale Weather Influences on
    Community Air Pollution Potential in the United States. —
    J. Air Pollut. Control Ass., Vol. 19, No. 4. 1969.

15. Lowrence, E.N.   High Values of Atmospheric Pollution in
    Summer of Kew and the Associated Weather. — Atmospheric
    Environment, Vol. 3, No. 12. 1969.

16. Noack, R.   Untersuchungen über Zusammenhänge zwischen
    Luftverunreinigung und meteorologischen Faktoren. —
    Angew. Met., Vol. 4, pp. 8—10. 1963.

THE STATISTICAL DETERMINATION OF MEAN AND MAXIMUM VALUES OF CONTAMINANT CONCENTRATION

E. Yu. Bezuglaya

When planning air-pollution observations, one of the necessary conditions is the possibility of comparing the results from different stations located in the same town or in different towns. In the hydrometeorological service this is to some extent ensured by carrying out regular observations, by standard methods of sampling, and sample analysis. We distinguish between stationary and mobile observations at different points of the town, and also observations under the jets of the industrial plants at different distances from the discharge sources. At stationary points samples are taken 2–3 times daily, i. e., 500 to 1,000 times yearly. Mobile observations are carried out once daily, or less often, and their total number over the year does not usually exceed 100–150. Observations under the jet of discharge sources are carried out episodically. At a given distance from the plant the number of measurements is rarely higher than 20–30. In some towns observations of air pollution are made during the period of investigation of the chemical composition of the air basin. only. The number of samplings and the analysis of the air samples may be different in this case. As a result, the values of the mean and maximum concentrations for any observation period are not comparable, due to differences in the number of measurements.

It was noted in /1/ that in many cases the distribution of the probability of contaminant concentration in city air is described satisfactorily by the logarithmic normal distribution law

$$f(q) = \frac{1}{sq\sqrt{2\pi}}\, e^{-\frac{\ln^2 \frac{q}{m}}{2s^2}}, \tag{1}$$

where $f(q)$ is the distribution density of the concentration of contaminant q; a and m are the parameters of the logarithmic normal distribution.

If the distributions of the quantities studied follow the logarithmic normal law, from (1) we can obtain analytical expressions for

140

the mean value q of the contaminant concentration, its variance σ^2, the coefficient of variation V, the maximum concentration of the contaminant q_M, with any exceedance probability, and for the probability F the appearance of a value q higher than any given q_M:

$$\overline{q} = me^{s^2/2}, \tag{2}$$

$$\sigma^2 = m^2 e^{s^2}(e^{s^2} - 1), \tag{3}$$

$$v = \frac{\sigma}{\overline{q}} = \sqrt{e^{s^2} - 1}, \tag{4}$$

$$q_M = me^{zs\sqrt{2}}, \tag{5}$$

$$F(q > q_M) = \frac{1}{2}\left[1 - \operatorname{erf}\left(\frac{\ln\frac{q_M}{m}}{s\sqrt{2}}\right)\right], \tag{6}$$

where $\quad z = \dfrac{\ln\frac{q_M}{m}}{s\sqrt{2}};$

for the probability of exceeding q_M in 0.1% of the cases

$$q_M = me^{3s}. \tag{7}$$

By means of these expressions we can obtain all the necessary characteristics from the experimental data. The calculations are however, very difficult in practice. Usually, the determination of q_M with different exceedance probabilities is of greatest interest. It can be obtained from (2) and (3) in terms of $\overline{q}$ and σ^2 :

$$q_M = \frac{\overline{q}}{\sqrt{1 + \frac{\sigma^2}{q^2}}} e^{z\sqrt{2}\sqrt{\ln\left(1 + \frac{\sigma^2}{q^2}\right)}}. \tag{8}$$

The calculation of $\overline{q}$ and σ^2 from the data of observations is not particularly difficult.

To facilitate the calculations of q_M, we plotted a nomogram by means of which q_M can be determined from known values of $\overline{q}$ for a given exceedance probability. Such a nomogram is given in Figure 1 for $F(q>q_M) = 0.1\%$. The probability of exceeding q_M in 0.1% of the cases is the most acceptable characteristic, since it agrees best with the observed value of the maximum in a large number of observations. If the measurements are carried out thrice daily, i. e., more than 1,000 times yearly, then 0.1% corresponds to one of them.

To test the possibility of determining the maximum concentration of the contaminant with the aid of the nomogram, we used the results of observations on sulfur dioxide, carbon monoxide, and nitrogen dioxide in 1968—1969 at 30 towns of the USSR (80 stationary points). First, the applicability of the logarithmic normal law to the data of the distribution of concentrations of contaminants was checked. As a rule, the concentrations of all the contaminants are distributed according to the logarithmic normal distribution law.

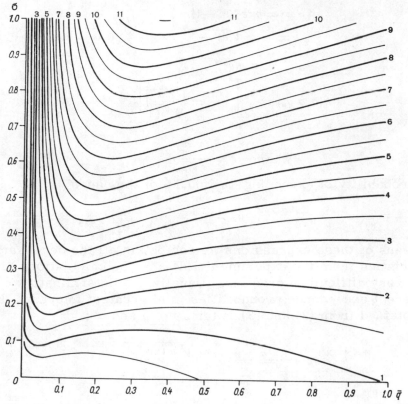

FIGURE 1. Nomogram for calculating q_M from given $\bar{q}$ and σ.

From the calculated mean contaminant concentrations and σ, we used the nomogram to determine the values of q_M with an exceedance probability in 1 and 0.1% of the cases. An analysis of the results showed that the maximum contaminant concentration, calculated for $F(q > q_M) = 1\%$, is usually lower than the observed value. Figure 2 gives the correlation between the calculated and observed maximum values for $F = 0.1\%$.

It is seen from the figure that the calculated maximum concentrations of the contaminants are scarcely ever lower than the observed values. The observed maximum usually corresponds to a 0.1—0.5% exceedance probability, since the number of measurements during the given period can be arbitrary, and therefore the

value of the recorded maximum may vary over wide limits.   By
the computing method, we are able to accurately determine the
maximum, with any exceedance probability.

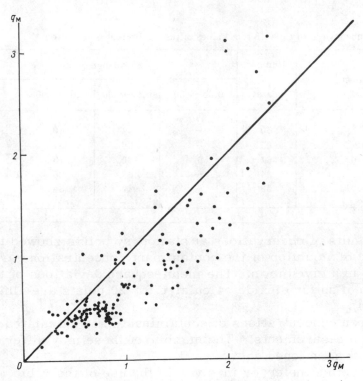

FIGURE 2. Correlation between the calculated (x-axis) and experimental (y-axis) maximum concentrations of the contaminants

From (8) it can be seen that the relative error in the deter-
mination of q_M depends on the error in the calculation of $\overline{q}$ and σ.
The data of the variance in the concentration of contaminants were
the analyzed, and it was found that variance depends on the type of
contaminant /4/.   The results indicated that σ depends mainly on
the value of $\overline{q}$, a characteristic feature of the logarithmic normal
distribution as the coefficient of variation is near to 1, and not on the
type of contaminant.   This means that the higher the mean level of
air contamination by the given contaminant, the higher is its
variance.   At the same time, a dependence on the type of contaminant is also observed.
    Table 1 shows that in most towns the coefficient of variation for
the carbon monoxide concentration varies between 0.5 and 1, and
exceeds 1 in a few cases only.   The mean-square deviation of the

concentrations of carbon dioxide, dust, and nitrogen dioxide, are between 0.5 and twice the mean value. The coefficient of variation for the concentration of soot varies within the same limits, but is usually greater than 1.

TABLE 1. Frequency (%) of the different values of the coefficient of variation V

Contaminant	Number of cases	Range of V			
		0.00–0.50	0.51–1.00	1.01–2.00	2.01
Sulfur dioxide	60	–	43	40	17
Nitrogen dioxide	65	–	60	34	6
Carbon monoxide	61	6	80	13	–
Dust	34	–	68	26	6
Soot	33	–	12	64	24

The results of observations at stationary points showed that the coefficient of variation of the contaminant concentration varies negligibly in a given town. The mean-square deviations of the concentration of sulfur dioxide calculated for 20 points are almost the same /3, 4/.

The mean concentrations of contaminants are calculated from the results of measurements. The number of these may differ considerably as mentioned above.

To determine the error involved in the use of the value $\bar{q}$ obtained from a limited selection, instead of the true parameter x, we use the expression for the confidence coefficient

$$p(\bar{q} - \bar{x}) < t_g \frac{\sigma}{\sqrt{n}} = a, \tag{9}$$

where t_g is the Student parameter, a the specified confidence coefficient

$$\bar{x} = \bar{q} \pm t_g \frac{\sigma}{\sqrt{n}}. \tag{10}$$

For a confidence coefficient of 0.95, $t_g = 1.96$. It was shown that for a contaminant concentration $\bar{q} = \sigma$. Therefore, to ensure a calculation of $\bar{q}$ with an error of 20%, the number of measurements n must not be less than 100. For $2q = \sigma$, the number of measurements must be increased to 400, and for $\sigma = 0.5q$ n must be reduced to 25.

It is known that because of the connection between adjacent terms of the sampling, the volume of information necessary for calculating $\bar{q}$ with a given accuracy is increased by the factor $\sqrt{\frac{1 + \rho(1)}{1 - \rho(1)}}$, where ρ is the value of the normalized correlation function in a time period equal to the interval between observations /2/. It is interesting to determine the form of the time-dependent correlation function and the corresponding value of the coefficient for different contaminants. Figure 3 shows the general form of the dependence of the normalized correlation function of air pollution on the time τ, calculated from observations of the concentrations of sulfur dioxide and nitrogen dioxide in Dnepropetrovsk in the winter of 1968. The time-dependent correlation for the concentrations of these contaminants is almost the same.

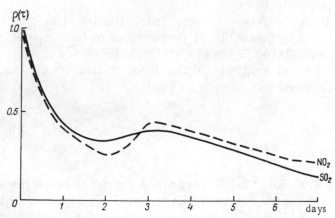

FIGURE 3. Time-dependent normalized $\rho(\tau)$ correlation functions of sulfur dioxide and nitrogen dioxide. Dnepropetrovsk.

A single concentration of a contaminant in the atmosphere was usually determined by taking samples during 20—30 min. In the system of the Hydrometeorological Service observations are carried out every three hours according to a sliding curve, three days weekly in the morning and three days in the afternoon. The time difference between evening and morning observations is 9 and 15 hours. Figure 3 shows that the correlation coefficient of air contamination is 0.80 when the difference in the times of measurement is less than 3 hours, 0.69 for a difference in time of 9 hours, and 0.51—0.55 for a difference in time of 15 hours. Thus, to increase the accuracy of calculation of the mean contaminant concentration we must increase the volume of necessary information for the given measurement periods by a factor of more than two.

In accordance with the above, we can make several recommendations for planning the work of the stations for observing the chemical composition of the atmosphere.

To ensure that the error in the calculation of the mean concentrations of sulfur dioxide, nitrogen dioxide, dust, and soot, will not
exceed 20%, the number of measurements for each of these contaminants must not be less than 200, and for $\sigma = 2q$ it must be
increased to 800.

Since the variance in the concentration of carbon monoxide is
lower than that of the other contaminants, 150—200 observations
suffice to achieve the required accuracy.

When organizing studies on the contamination of the city air
basin, or carrying out observations under the jet of industrial
enterprises, we must take into account the variation in the concentration during the year, and carry out such observations during
the period of possible maximum.

To ensure that the data characterizing the level of city air
contamination are comparable, it is convenient to use the maximum
concentration calculated from (8), and not the observed.

Besides, a critical analysis of the information on air pollution
is necessary to weed out doubtful results.

Bibliography

1. Bezuglaya, E. Yu. Ispol'zovanie statisticheskikh metodov
      dlya obrabotki dannykh nablyudenii za zagryazneniem
      vozdukha (Use of Statistical Methods for Processing the
      Results of Observations on Air Pollution). — Trudy GGO,
      No. 238, pp. 42—47. 1969.
2. Borisenkova, E. M. (Editor). Vvedenie v statisticheskie
      metody obrabotki gidrometeorologicheskoi informatsii na
      ETsVM (Introduction to Statistical Methods for Processing
      Hydrometeorological Information on Computers). —
      AANII. Leningrad. 1966.
3. Juda, J. Planung und Auswertung von Messungen der Verun-
      reinigungen in der Luft. — Staub, Vol. 28, No. 5, pp. 186—193.
      1968.
4. Stratmann, H. and D. Rosin. Untersuchungen über die
      Bedeutung einer empirischen Kenngrösse zur Beschreibung
      der Häufigkeitsverteilung von SO_2-Konzentrationen in der
      Atmosphäre. — Staub, Vol. 24, No. 12. 1964.

THE QUESTION OF THE SELECTION OF THE NUMBER OF SAMPLING STATIONS AND THE FREQUENCY OF OBSERVATIONS OF AIR POLLUTION

B. B. Goroshko

The taking of samples for studying atmospheric pollution and their chemical analysis need an appreciable time. The question of the optimum number of sampling stations and the sampling frequency are therefore of considerable importance.

The problem is solved easily if one single source is responsible for air pollution. We shall start with this particular case, and then pass to the question of the organization of observations in the territory of a city with a large number of sources of discharge of harmful substances.

The questions of the organization of studies on atmospheric pollutions in the vicinity of large plants have already been treated in part /8—11/. It was noted that the distance between observation stations must be shorter when the station is nearer to the polluting source. The distance between stations can be increased far from the source, since on the lee side much of the pollution is due to unorganized discharges and discharges from low stacks, and their effect decreases rapidly. Besides, the concentration of contaminants discharged from high stacks increases at first rapidly with distance from the source, passes through a maximum, and then decreases slowly /1, 4, 7, 17/. According to /1, 17/, the highest concentrations are observed at distances between 10 and $40H$, with a maximum at approximately $20H$, where H is the height of discharge of contaminants. Therefore, the observations must be more frequent at distances over this range. The need to carry out observations at remote stations depends on the power of the discharge, the height of the source, and the meteorological conditions of the dispersion of the contaminant. It can be determined by calculating the expected values of the concentrations, or experimentally from the results of trial measurements.

It is advisable to place 3—5 sampling stations at each range at a distance of 50—40 m from one another, depending on the width of the jet. Such an arrangement is necessary, since even in the case of a stable jet direction, its axis oscillates as a result of fluctuations in the wind vector. The jet expands and becomes blurred with increase in distance from the source. Thus, it is difficult to determine the direction of its axis, and to obtain the maximum axial concentrations we must increase the number of sampling points to at least 5 (the distance between them is up to 200 m, or even more at times).

Under the jet of a source with a stable discharge of harmful substances a diurnal variation in the concentrations is observed. This is caused by variation in the meteorological elements determining dispersion of the contaminants. It was found theoretically /3/ and by processing the experimental material /7/ that the value of the concentrations is appreciably affected by the stratification, the wind speed, and the degree of turbulence in the atmosphere. Recent studies indicate that there is a certain dangerous temperature gradient at which the maximum concentrations at ground level are observed.

TABLE 1. Variation in the mean monthly temperature in the 0.5—2 m layer in August

Weather	Observation period (hr)										
	0	4	6	8	10	12	14	16	18	20	22
Fair	-0.4	-0.7	-0.5	0	0.8	1.2	1.6	0.7	-0.8	-0.8	-0.3
Overcast					0.6	0.7	0.8	0.4			

It is interesting to consider the variation in the mean monthly temperature differences Δt in the 0.5—2 m layer. Table 1 gives the mean values of Δt for fair weather, when during the observation period the quantity of high-level and middle-level clouds did not exceed 7 scale points, and of low-level clouds 3—4 scale points, and for overcast weather. The table shows that the greatest variations are observed in the morning and evening in fair weather: from 8 to 10 hours from 0 to 0.8°C, and from 16 to 18 hours from 0.7 to -0.8°C. The considerable variation in Δt in the course of an hour shows that if only one observation is made each hour, the dangerous conditions might not be recorded. Therefore, sampling must be carried out at least once an hour. The meteorological observations must also be carried out hourly. It is very important to take more frequent observations in the period of possible greatest variation in the gradient, that is, in fair weather in daylight. In general, to study sources with a high discharge of harmful

substances in the summer period in the middle belt of the European
Territory of the Union it is necessary to take samples between 6
and 21 hours to record the most dangerous conditions with maximum
concentrations.

Under urban conditions the study of the chemical composition of
the air is particularly difficult, since the fluctuation in the concentration level at each moment of time depends on several factors, some
some constant and others that vary over wide ranges in the course
of a day or even an hour. Factors such as the town dimensions,
the local relief, the planning and building of residential areas, and
the micrometeorological features of the city, the existence of green
spaces and reservoirs, the location of the industrial sources, the
existence of organized and unorganized discharges, their height, and
the temperature of the gases evolved, belong to the first group.
Variation in the volume of harmful substances entering the atmosphere, fluctuation in the meteorological factors determining the
degree of dispersion of the contaminants, the chemical stability of
the elements and their physical properties, periodical and random
sources belong to the second group. All these factors operate in a
large industrial center and determine the degree of pollution of its
air basin.

As the air mass passes over the town territory, it will gradually
become saturated by harmful substances. Thus, the concentrations
will be lower on the windward side and higher on the leeward side.
We can give as an example the concentrations of sulfur dioxide
measured in the northern part of Leningrad. The maximum concentrations with northern winds were 3.5—4 times less than with
southern winds. The larger the city dimensions, the larger is the
time during which the same air mass remains above its territory,
and the higher are the concentrations.

The planning of the residential areas affects their aeration. If
the buildings are close together, the wind speed is decreased, or, in
other words, air exchange is slower, and stagnant zones are formed,
thus leading to the accumulation of harmful substances. This is
particularly dangerous in the case of low discharges. The contemporary tendency to have spaced buildings favors air exchange in the
residential area, and therefore a decrease in the concentrations /13/.

The microclimatic features of urban building include the
existence of an island of warmer air at the town center, and as a
result the creation of local air circulation directed toward the town
center /5, 14/. Local circulation leads to the transfer of harmful
substances from the periphery to the center of the town, and thus to
an increase in the concentrations there.

The arrangement of discharge sources in the city can be divided into two types /6/. In the first type the sources are concentrated in a single industrial area. It is clear that in this case the maximum concentrations will be observed in the direction of the air stream, i. e., in the direction of displacement of the principal jet. In the second type the discharge sources are scattered over the whole territory. In this case a general background of pollution of the air basin is created, fairly homogeneous over its entire area.

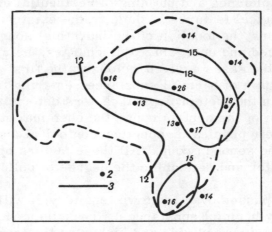

FIGURE 1. Distribution of the normalized mean concentrations of carbon monoxide over the town territory in the fall period:

1) city boundary; 2) sampling stations; 3) isolines of equal concentrations.

Figure 1 shows the mean concentrations of carbon monoxide in the the fall period (September—October) for one of the towns in the Ukraine. Since the industrial plants are distributed over the whole town, and since motor traffic discharges large quantities of carbon monoxide over the entire town also, the mean concentrations differ little over the territory. Such a classification of the towns is of fundamental importance, since the method for organizing the observations must depend on it.

The height of the discharge of contaminants has a considerable effect on the concentration field. All the sources can be divided arbitrarily into high, above 50 m, and low, below 50 m, and unorganized discharges. The concentrations increase in the zone of a high source jet with increase in turbulent exchange. A decrease in turbulent exchange during inversion temperature distribution and low wind speed must lead to an increase in ground level concentrations with low and unorganized discharges of contaminants.

A number of authors /18, 22, etc./ note that the maximum concentrations are observed during calm or near-calm wind speeds, while a stronger wind leads to a purification of the air basin. Such conclusions are correct for low discharges. This is confirmed in /15/, which gives the dependence of the concentrations of dust and sulfur dioxide on the wind speed in the warm and cold periods. The two concentration maxima observed during calm, and at wind speeds of 4—6 m/sec, can be explained physically by assuming that in the first case the increase in the concentration was caused by the low sources, and in the second case by the high sources. A wind speed of 4—6 m/sec was found to be dangerous for high sources located within the city limits. This wind speed caused the appearance of a second maximum. In other cases the second maximum may be missing. It was shown theoretically /1, 2/, and by processing the experimental data /7/, that the concentrations in the region of high sources decrease with the establishment of an inversion distribution of temperature, and with a weak wind in the ground layer. In town, in the presence of low sources, these conditions will lead to an increase in the concentrations as a result of a decrease in air exchange.

Of the other factors that determine the degree of contamination of a town, we must consider the fluctuations in the volume of discharge of contaminants into the atmosphere. These fluctuations depend on the manufacturing processes and on the work periodicity. Small plants with a low discharge generally operate during the first half of the day. The exhausts of motor vehicles have their specific features. They are transported and distributed over the whole city, and the highest exhaust intensity occurs also during the first half of the day. The combustion of fuel for heating is most intense in the morning and evening. This process also passes through a yearly cycle, with a maximum in the winter months.

These facts complicate the development of strict principles for carrying out studies on the pollution of urban air. This problem is solved differently in different countries. Stations with round-the-clock recording of the main factors are organized in the USA, usually with one such station in each city /19/. Naturally, with such an organization of the studies, the results cannot characterize the pollution of the atmosphere of the whole town. In Nashville (USA) /23/ a method was used for a gradual approach to the optimum number of stations for controlling the purity of air in the atmosphere. In the initial period, 119 sampling stations were organized over the town territory with an area of 29 sq. miles. It was later shown

that the samples from a number of stations gave identical results,
and the redundant stations were closed down. In this way the
required optimum number of sampling stations was found. With
this procedure it is possible to find the most representative
sampling locations for a given town region, but the method is
laborious and expensive.

In other cases, the organization of the observations, the selection
of the number of sampling stations, the sampling frequency, and the
total number of samples taken, are based on the fact that the concentration, in fact, follows the logarithmic normal distribution law,
independently of the length of the series of observations (week,
month, year) and the type of contaminant /20/. The total frequency
is then given by the following expression:

$$F(Si) = \frac{1}{\sigma_1 \sqrt{2\pi}} \int_{-\infty}^{\lg Si} \exp\left[\frac{-(\lg S - a)^2}{2\sigma_1^2}\right] d(\lg S),$$

where $F(Si)$ is the probability that the contaminant concentration S
is less than Si; a is the mean value of the logarithm of the concentration; σ_1 is the standard deviation of the concentrations.

It follows that to obtain the basic characteristics of the pollution
(mean concentrations, maximum and frequency) we must carry out
systematic measurements on a dense network of stations (at the
apexes of a 2×2 km square). The duration of the measurements is
between 20 and 30 minutes. To calculate the mean concentration
accurate to within 20%, we must carry out no less than 60 different
measurements, and the duration of the observation period must be
not less than a year. The day and hour of the individual observations should be selected by means of tables of random numbers.

The main shortcoming of such an organization of the observations is that it gives characteristics of the town atmosphere as
a whole, and does not enable us to determine regions of maximum
and minimum pollution. Besides, a rigid specification of the
location of the sampling stations (deviation from the corresponding
network node not greater than 200 m) may mean that the stations
may become nonrepresentative and in some cases this specification
is simply not possible under urban conditions.

In other cases a geometric network of sampling stations was
used to organize the observations /21/ (Detroit, Windsor, Paris,
Nashville). In Los Angeles the characteristic local trajectories
of the air masses were taken into account.

The fundamental principles for organizing observations in the
USSR have been briefly described in /12/, and in greater detail in
/16/. It is indicated that the investigations are based on samples

taken at stations set up with a frequency of one station per
10—20 km^2 area in flat terrain, and per 5—10 km^2 area in broken
terrain.   Thus, in the first case the distance between stations will
be roughly 3.2 to 4.5 km, and in the second case 2.2 to 3.2 km.   With
such a frequency of the stations it is possible to determine the air
pollution characteristics of microregions and the town as a whole.
We believe that the use of such a network of stations has considerable advantages.   Since the distance between stations is not specified rigidly, they can be located in a more efficient manner.   This
is true in particular with regard to the selection of the more representative points.   Besides, it is possible to obtain a detailed
characteristic in the most densely populated and most polluted
regions by organizing a dense network of sampling stations in such
places and reducing their number in the suburbs, where the pollution
is a minimum.

If the town is built on rough terrain, it is necessary to increase
the number of stations due to the particular features of the relief,
which favor air stagnation or the appearance of local currents.
Sampling stations must be located at both low and high points of the
relief.

To test this reasoning, a town extending for 25—30 km in the
N-S direction and for 20 km in the W-E direction was studied.
Twenty nine sampling stations were set up over its territory.   If
their distribution had been regular, each one would have corresponded to an area of 20 km^2; however, the positions of discharge
sources, residential areas, relief, etc., were taken into account, and
thus in the most polluted areas there was one station per 5—10 km^2,
with a less dense network in the suburbs.   Subsequently, all
29 stations were transferred to the northern region of the town,
with a territory of about $^1/_5$ of the total area of the town, where for-
merly there were 7 sampling stations.   When the dense network of
sampling stations was set up, the distance between them was 1—2 km.

From the results the mean monthly concentrations, and the frequencies of concentrations higher than certain values were calculated, and isolines of equal monthly concentrations were plotted for
the following components; dust, sulfur dioxide, phenols, nitrogen
dioxide, carbon monoxide.   An analysis of the results shows that the
increase in the number of sampling stations did not lead to an
increase in the accuracy of the measurements of the fundamental
characteristics in time or space.

Table 2 gives the distance between the stations and the deviation
from the mean monthly and maximum values of the concentrations
in percent for two towns in the Ukraine.   It is seen that the distance
between stations under urban conditions is not the determining

factor leading to difference in the values of the concentrations. Thus, the distance between the stations (station 1) 1—2, 7—8, 6—8, 5—9, varies between 2.6 and 3.6 km, and in the second town for distances between the stations of 1.2 to 2.6 km, the differences in the concentrations do not exceed 40%, which is within the experimental error of the method. In other cases, considerable differences between the mean concentrations, and, in particular, between their maximum values, are observed even for smaller distances.

TABLE 2. Excess above the mean monthly and maximum concentrations (in percent) in the towns for different distances between the stations

| No. of stations | Distance between stations (km) | Excess concentrations | | | | | |
| | | mean | | | maximum | | |
		sulfur dioxide	nitrogen dioxide	phenol	sulfur dioxide	nitrogen dioxide	phenol
		Station 1					
1.2	3.6	0	8	13	12	3	40
2.10	3.0	14	20	20	73	3	65
3.4	2.6	0	10	44	13	18	67
4.10	3.6	32	40	42	86	6	65
5.12	1.4	22	55	2	23	18	60
6.11	1.6	33	74	9	41	71	24
7.8	2.6	33	20	4	26	32	8
6.8	2.8	22	11	12	28	18	11
5.9	3.4	17	30	29	20	13	47
10.12	2.2	18	32	5	50	38	36
5.11	2.0	9	72	23	7	75	60
4.12	3.4	17	59	44	59	41	45
		Station 2					
2.12	1.2	2	12	4	18	19	9
3.12	2.4	7	10	7	16	18	20
4.5	1.2	11	12	3	40	0	89
5.6	1.5	10	9	15	8	20	85
6.9	1.2	7	23	10	33	3	30
7.8	1.2	10	2	3	8	16	2
8.9	1.5	16	16	0	37	10	13
9.13	1.5	18	16	46	18	15	31
10.11	1.1	8	0	0	35	63	6
11.12	1.8	2	5	13	7	43	3
12.13	3.0	33	10	54	29	23	23
6.13	0.8	24	7	27	10	13	2
2.14	2.6	18	7	22	0	9	23

A detailed investigation of the causes of variation in the concentrations between points showed that the difference is determined
mainly by the location of the sampling stations and less by the
distance between them.   The vicinity of busy highways and industrial areas was the most important factor here.   Therefore,
a network density with a distance of 3—4 km between stations is
sufficient for obtaining the general characteristics of the field of
concentrations over the town territory.   If the concentration of
contaminants is determined mainly by the influence of local sources,
it is expedient to use automatic sampling machines set up at
different distances from the center of discharge.

The determination of the necessary sampling frequency and of
the total duration of the studies is also of great importance.

Since the meteorological factors and the discharge of contaminants vary in the course of the year, it is obviously necessary to
carry out observations of the concentration of contaminants over
as wide as possible diurnal fluctuations.   It can be assumed that
these fluctuations will be maximum during daylight.

It is known that the meteorological factors and the discharge
into the atmosphere vary most in the daytime.   These questions
were discussed in /5, 6/, and the dependence of the concentration
on the variations in the meteorological factors during the day and
year, and on the height of discharge was shown.   From the results
it was found that samples must be taken over the whole day.   More
detailed observations should be carried out between 6 and 21 hours.
In analogy with the meteorological observations, an interval of
3 hours between samplings is proposed.   To reduce the number of
observers, it is suggested that a sliding curve be used, that is,
observations be taken three days weekly in the morning and three
days in the afternoon, alternately.   Naturally, if such a curve of
observations is used the maximum values will not always be recorded.   This can be seen from the results given in Table 3, which is
based on hourly 20-min long samplings of sulfur dioxide at one of
the city stations in August 1969.   The results in the second column
characterize the diurnal maximum concentrations from all the
series of observations; the third column gives the values corresponding to daylight (at 7, 10, 13, 14, 17, and 20 hours); the fourth
column shows the deviation in the concentrations of the third
column from the second one; the fifth column the values when
samples are taken according to the sliding curve (one day at 7, 10,
and 13 hours, the second one at 14, 17, and 20 hours); the seventh,
eighth and tenth columns give the mean concentrations corresponding to all the periods, in daylight and according to a sliding
curve, and the ninth and eleventh the deviation from the means for
all the periods.

TABLE 3. Maximum and mean normalized concentrations of sulfur dioxide (mg/m^3) and their deviation at different sampling frequencies (Station 1)

No.	Maximum of all periods	Maximum of periods 7, 10, 13, 14, 17, 20	Deviation (%)	Maximum according to sliding curve	Deviation (%)	Mean diurnal of all periods	Mean diurnal of periods 7, 10, 13, 14, 17, 20	Deviation (%)	Mean diurnal according to sliding curve	Deviation (%)
1	2	3	4	5	6	7	8	9	10	11
1	0.56	0.50	11	0.50	11	0.29	0.29	0	0.37	-28
2	0.25	0.23	8	0.23	8	0.18	0.18	0	0.19	- 6
3	1.01	1.01	0	1.01	0	0.90	0.94	- 4	0.94	- 4
4	0.61	–	–	–	–	0.60	–	–	–	–
5	1.34	1.34	0	0.40	70	0.45	0.53	-18	0.27	40
6	0.52	0.33	37	0.33	37	0.25	0.19	24	0.21	16
7	0.78	0.78	0	0.78	0	0.46	0.46	0	0.46	0
8	0.13	0.13	0	0.12	8	0.09	0.08	11	0.06	33
9	0.46	0.46	0	0.46	0	0.27	0.31	-15	0.31	-15
10	0.23	0.10	56	0.10	56	0.11	0.07	36	0.03	73
11	0.28	0.17	39	0.17	39	0.09	0.07	22	0.06	33
12	0.35	0.35	0	0.31	11	0.22	0.25	-14	0.24	- 9
13	0.66	0.53	20	0.53	20	0.44	0.46	- 5	0.49	-11
14	0.43	0.43	0	0.43	0	0.32	0.34	- 6	0.35	- 9
15	0.28	0.28	0	0.23	18	0.19	0.20	- 5	0.17	10
16	0.68	0.51	25	0.51	25	0.26	0.22	15	0.26	0
17	0.38	0.38	0	0.28	26	0.21	0.20	5	0.16	24
18	0.34	0.33	3	0.33	3	0.17	0.15	12	0.18	- 6
19	0.28	0.28	0	–	–	0.12	0.21	-75	–	–
20	0.24	0.24	0	0.10	58	0.09	0.10	-11	0.03	67
21	0.52	0.51	2	0.47	10	0.35	0.36	- 3	0.30	14
22	0.33	0.33	0	0.21	36	0.16	0.17	- 6	0.14	12
23	0.69	0.69	0	0.23	67	0.11	0.19	-73	0.13	-18
24	0.23	0.17	26	0.17	26	0.12	0.09	25	0.09	25
For the month	1.34	1.34	0	1.01	25	0.25	0.25	0	0.24	4

It follows from the given table that by choosing the maximum values from the monthly data and calculating the mean monthly values of the concentrations, the degree of pollution can be satisfactorily characterized when sampling is performed according to the sliding curve. The deviation of the maximum and mean monthly values of the concentrations when sampling is performed six times daily and according to the sliding curve from the values obtained by hourly sampling does not exceed the experimental error in the determination of the given component. A poorer agreement between the maximum and mean daily concentrations is observed at different sampling frequencies. However, a detailed examination of the table shows that the maximum deviation percent is observed at low values of the concentrations, when the error of determination increases. Similar results were also found at two other stations in the same town, and when samples of sulfur dioxide, nitrogen oxides, phenol, carbon monoxide, and dust, were taken at all three stations.

The results indicate that the concentration field of the different components shows a certain time lag. Therefore, a sliding curve of air sampling is recommended for determining the basic charac-teristics of atmospheric pollution over a town.

To decrease the danger of missing a maximum concentration, it is desirable to use automatic devices to take samples under the jets of the main pollution sources.

Bibliography

1. Berlyand, M. E. and R. I. Onikul. Fizicheskie osnovy rascheta rasseivaniya v atmosfere promyshlennykh vybrosov (Physical Principles for Calculating the Dis-persion of Industrial Discharges in the Atmosphere). – Trudy GGO, No. 234, pp. 3–27. 1968.
2. Berlyand, M. E., E. L. Genikhovich, and R. I. Onikul. O raschete zagryazneniya atmosfery vybrosami iz dymo-vykh trub elektrostantsii (Calculation of Air Pollution from the Stacks of Power Plants). – Trudy GGO, No. 158. 1964.
3. Berlyand, M. E. et al. Chislennoe issledovanie atmosfernoi diffuzii pri normal'nykh i anomal'nykh usloviyakh stratifi-katsii (Numerical Study of Atmospheric Diffusion under Normal and Anomalous Stratification Conditions). – Trudy GGO, No. 158. 1964.
4. Berlyand, M. E., E. L. Genikhovich, and O. I. Kurenbin. Vliyanie rel'efa mestnosti na rasprostranenie primesei ot istochnikov (Influence of Local Relief on the Propagation of Contaminants from Sources). – Trudy GGO, No. 234, pp. 28–44. 1968.

5. Burenin, N. S. and B. B. Goroshko. K izucheniyu zagryazneniya atmosfery goroda promyshlennymi vybrosami
   (Study of the Pollution of the Atmosphere of a City by
   Industrial Discharges). — Trudy GGO, No. 238, pp. 136—144.
   1969.
6. Burenin, N. S., B. B. Goroshko, and B. N. P'yantsev.
   Ekspeditsionnoe izuchenie zagryazneniya vozdushnogo
   basseina promyshlennykh gorodov (Study of an Expedition
   on the Pollution of the Air Basin of Industrial Cities). —
   Trudy GGO, No. 234, pp. 100—108. 1968.
7. Goroshko, B. B. Nekotorye osobennosti rasprostraneniya
   vrednykh primesei ot vysokikh istochnikov v zavisimosti
   ot sinoptiko-meteorologicheskikh faktorov (Some Features
   of the Propagation of Harmful Contaminants from High
   Sources as a Function of the Synoptic-Meteorological
   Factors). — Trudy GGO, No. 207. 1968.
8. Goroshko, B. B. Postanovka eksperimental'nykh rabot po
   izucheniyu rasprostraneniya vrednykh primesei ot moshchnykh istochnikov (Organization of the Experimental Studies
   on the Propagation of Harmful Contaminants from Powerful Sources). — Trudy GGO, No. 234, pp. 109—115. 1968.
9. Goroshko, B. B. et al. Meteorologicheskie nablyudeniya pri
   issledovanii promyshlennykh zagryaznenii prizemnogo
   sloya vozdukha (Meteorological Observations during the
   Study of the Industrial Pollution of the Ground Air Layer). —
   Trudy GGO, No. 138. 1963.
10. Goroshko, B. B., A. S. Zaitsev, and V. Ya. Nazarenko.
    Voprosy metodiki i rezul'taty issledovaniya zagryazneniya
    atmosfery s pomoshch'yu vertoleta (Problems of the
    Methodology and Results of the Study of Atmospheric
    Pollution by Helicopters). — Trudy GGO, No. 234, pp. 85—94.
    1968.
11. Goroshko, B. B. Ekspeditsionnoe issledovanie zagryazneniya
    promyshlennymi vybrosami gorodov (Study of the Expe-
    dition on the Pollution of a City by Industrial Discharges). —
    In: Meteorologicheskie aspekty promyshlennykh zagryaz-
    nenii atmosfery. Moskva, Gidrometeoizdat. 1968.
12. Goroshko, B. B. and T. A. Ogneva. Osnovnye printsipy or-
    ganizatsii obsledovaniya sostoyaniya zagryazneniya at-
    mosfery v gorodakh (Basic Principles for Organizing the
    Study of the State of Pollution of the Atmosphere in Towns).—
    Trudy GGO, No. 238, pp. 123—135. 1969.
13. Kolyuzhnyi, D. N., V. I. Pal'gov, and Yu. V. Dumanskii.
    Vliyanie kharaktera gorodskoi zastroiki na nazemnye inso-
    lyatsii i aeratsii v usloviyakh UkrSSR (Influence of the

Character of City Buildings on Ground Insolation and
Aeration under the Conditions of Ukrainian SSR).— In:
Voprosy prikladnoi klimatologii. Leningrad, Gidrome-
teoizdat. 1960.

14. Rastorgueva, G. P.   Osobennosti termicheskogo rezhima
gorodov (Characteristics of the Thermal Regime of
Cities).— Trudy GGO, No. 238, pp. 145—152. 1969.

15. Son'kin, L. R., E. A. Razbegaeva, and K. M. Terekhova.
K voprosu o meteorologicheskoi obuslovlennosti zagryaz-
neniya vozdukha nad gorodami (The Problem of the
Meteorological Causes of Air Pollution above Cities).—
Trudy GGO, No. 185. 1966.

16. Metodicheskie ukazaniya Upravleniyam gidrometeorologicheskoi
sluzhby po obsledovaniyu zagryazneniya vozdukha v gorodakh i promyshlennykh tsentrakh (Instructions for
Operating the Hydrometeorological Service for the Study
of Air Pollution in Towns and Industrial Centers).—GGO.
1970.

17. Ukazaniya po raschetu rasseivaniya v atmosfere vrednykh
veshchestv (pyli i sernistogo gaza), soderzhashchikhsya v
vybrosakh promyshlennykh predpriyatii (Instructions
Regarding the Calculation of the Diffusion in the Atmosphere of Harmful Substances (Dust and Sulfur Dioxide)
Contained in the Discharges of Industrial Plants).
CH-369-67. Leningrad, Gidrometeoizdat. 1967.

18. Dickson, R. R.   Meteorological Factors Affecting Particulate
Air Pollution of a City.— BAMS, Vol. 42, No. 8. 1961.

19. Gräfe, K.   Lufthygienische Probleme in Pennsylvania, New
Jersey and New York.— Städtehygiene, Vol. 19, No. 4. 1968.

20. Juda, J.   Planung und Auswertung von Messungen der Verun-
reinigungen in der Luft.— Staub Reinhaltung der Luft,
No. 5. 1968.

21. Katz, M.   Measurement of Air Pollutants. World Health
Organization. Geneva. 1969.

22. Lawrence, N. E.   Forecasting Air Pollution Potential.—
Mon. Weath. Rev., Vol. 88, No. 3. 1960.

23. Stalker, W. W. and R. C. Dickerson.   Sampling Station and
Time Requirements for Urban Air Pollution Surveys.
Pt. 3. Two- and Four-Hour Soiling Index.— J. Air
Pollut. Control Ass., Vol. 12, No. 4. 1964.

ANNUAL AND DIURNAL VARIATION IN THE CONTENT OF ATMOSPHERIC CONTAMINANTS UNDER URBAN CONDITIONS

E. Yu. Bezuglaya, A. A. Gorchiev,
E. A. Razbegaeva

The level of the content of contaminants in the air of cities shows considerable fluctuations. These fluctuations are a function of the meteorological factors and the discharge parameters, which in turn do not remain constant.

The meteorological factors are characterized by two basic cycles of variation, diurnal and annual, while the discharges have three cycles: diurnal, annual, and weekly. The interaction of these cycles yields a certain diurnal and annual variation in the content of contaminants in the atmosphere. The study of this is very important when planning the work regime of industrial plants with effective allowance for the meteorological factors.

A study of the annual variation based on observations in many towns of the USA, Canada, and Western Europe showed the existence of a maximum concentration, due mainly to higher discharges of contaminants when there is increased consumption of fuel for heating purposes /9—11/. Barrett /8/ indicates that the annual fluctuations in the dust are similar over the whole of England and have a high correlation. Seasonal characteristics of the development of atmospheric processes have a great influence on the annual variation in the pollution of the ground air layer. Thus, in Tokyo the content of sulfur dioxide and nitrogen oxides in summer is half that in winter, due to the abundant monsoon rains in summer /12/. The increased air pollution and the photochemical smog in Los Angeles during the end of the summer and autumn are clearly connected with the greater length of sunny periods, the continual inversions of the precipitation, and the breezy winds /9/. Measurements of the settling dust /6/, the concentrations of sulfur dioxide and dust /5/ in the air, and also of 3, 4-benzopyrene /7/ in Leningrad, showed the existence of a summer maximum in the different years.

From an analysis of the observations in a number of towns of the USSR Son'kin /5/ established some characteristics of the annual variation in the dust content of the air in the different geographical

regions.   In particular, a summer maximum of the contaminant
concentration was observed in Western Siberia and in Kazakhstan, and
and a winter maximum in Eastern Siberia.

Studies are being carried out on the diurnal variations in the air
pollution of cities. A maximum of the concentrations in the morning hours and a minimum after midday have been observed in many
countries /10, 11/. Studies in Montreal showed the existence of two
types of diurnal variations, depending on the season: the amplitude
of the fluctuations is small in the winter and large in summer, with
a gradual change from the first type to the second in spring and an
abrupt change in autumn /13/.

The contaminants with highest concentrations in the air of cities
are sulfur and nitrogen oxides, carbon monoxide, and dust, and
therefore it is the diurnal and annual variation in these contaminants
which is the most interesting. Carbon monoxide and nitrogen
oxides are formed mainly by motor vehicles, whose number increases from year to year, and also by the chemical, oil-processing,
metallurgical, and wood-processing industries. Sulfur dioxide is
contained in the exhausts of hydroelectric and thermoelectric
power plants, and other sources of air pollution discharging at
heights between 10 and 200 m. Dust is produced by cement plants
and thermoelectric and hydroelectric power plants.

We give below the results of a study on the diurnal and annual
variations in the concentrations of these contaminants, obtained from
an analysis of the results of systematic measurements at stationary
points in 50 towns of the country in 1968—1969.

The annual variation is analyzed from observations in towns
with several sampling stations at which the periods and the number
of measurements coincided roughly. The mean values for each
month were calculated (q_{av}), fixed with reference to the mean annual
concentrations of the contaminant in the town air (q_{an})

$$M = q_{av}/q_{an}.$$

With such an approach it is possible to find the months with the
highest deviations from the mean value, and to compare the seasonal
variations in the air pollution level in the different towns. The
mean monthly values M were used to analyze the annual variations
in the concentration of contaminants.

A considerable increase in the concentration of soot was observed during the cold season. It is caused by the switchover of
the industrial plants to winter regime.

The seasonal variations in the dust content of the air are also
affected by the snow or vegetation cover of the soil. The absence

of a developed vegetation cover after the melting of snow is the
main reason for the spring maximum of the dust content observed
in many regions of the country. In 1969, a winter (February)
maximum of the dust concentration, caused by dust storms, was
observed in many towns of the Ukraine, Lower-Volga area, and the
Northern Caucasus.

The annual variation in the gaseous contaminants is more
complex, due to the different effect of the meteorological conditions
on the propagation of contaminants from high and low sources. The
distribution of industrial plants over the town territory, the
difference in the character and height of their discharges, play an
important part in the creation of annual variation in the content of
contaminants in the atmosphere.

In towns where low discharges (such as those of motor vehicles)
are the main source of contaminants, the variation in the content
of contaminants is relatively smooth. But in the presence of high
discharges the influence of the meteorological conditions affecting
the dispersion of contaminants will not be well-defined, and the
determination of the annual variation of the content of contaminants
involves some difficulties. The operation of boiler-rooms in winter
only, the use of fuels with a different content of sulfur combinations
over the year, the variation in the operation of the plants, lead to
high fluctuations in the amount of discharges of sulfur dioxide, and to
complex annual variations in the concentrations. As a result, in
some towns the seasonal variations are clearly marked, while in
others fluctuations connected with the character of the discharges
are superposed on the well-defined basic variation.

In spite of these difficulties, the analysis of the mean monthly
values of M for gaseous contaminants led to the determination of
four basic types of seasonal variations, represented in Figure 1.

I. Higher content of contaminants in spring and autumn, lower
content in winter and summer.

II. Higher content in winter and summer, lower content in spring
and autumn.

III. Annual variation with maximum in winter and minimum in
summer.

IV. Annual variation with maximum in summer and minimum in
winter.

The analysis shows that although different contaminants do not
vary in exactly the same way during the year, their maximum concentrations are observed in the same seasons.

The basic type of annual variation was established for each town
separately for 1968 and for 1969. The map in Figure 2 indicates
the regions in which the different types predominate.

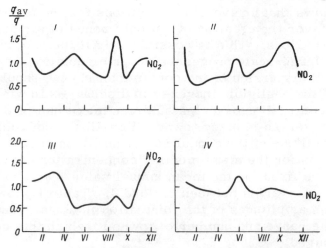

FIGURE 1. Types of annual variation in the concentrations of contaminants:

I) Leningrad; II) Syktyvkar; III) Irkutsk; IV) Nikolayev.

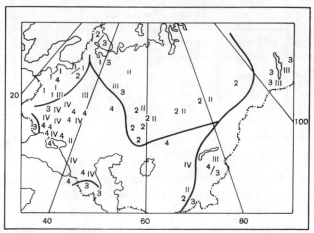

FIGURE 2. Distribution of the types of annual variation in air pollution in 1968 (Roman figures) and 1969 (Arabic figures)

A winter maximum of the concentration of contaminants in the air of cities is mainly observed in the towns of Eastern Siberia, Yakut, the Far East, Middle Asia, and some towns of Transcaucasia, and a summer maximum above most of the European Territory of the USSR, Ukraine, and Northern Caucasus. Two yearly maxima (in winter and summer) were observed in the towns of the Ural, the northeast of the European Territory of the USSR, and Western Siberia. Two annual maxima in spring and autumn were observed in some towns of the Baltic region, and in the northwest of the European Territory of the USSR.

Figure 2 shows that unfavorable conditions for the propagation of discharges favor the creation of not only zones of air pollution, but even large regions.   This is possible only if the propagation is determined by large-scale processes of air circulation, ensuring the transfer and dispersion of the contaminants.   We should note that in spite of the negligible increase in discharges in winter caused by the combustion of fuel, a summer maximum of the pollution was observed in many towns.   For all the contaminants studied, the amplitude of the annual variation (A) in each town was found and then, as for the mean monthly concentrations of contaminants, these were fixed for the mean annual value $A/q_{av.an}$.   It was found that $A/q_{av.an}$ varies between 0.2 and 4.   Table 1 shows the frequency of the amplitudes of the fluctuation in contaminants fixed according to the range and number of towns in which this contaminant was measured.

TABLE 1. Frequency % of the different amplitudes of the annual fluctuations in the concentration of contaminants

Contaminant	Range of amplitude				Number of cases
	0.00–0.50	0.51–1.00	1.01–2.00	> 2.00	
Dust	8	38	32	22	37
Sulfur dioxide	2	31	49	18	84
Nitrogen dioxide	8	27	49	16	85
Carbon monoxide	19	62	19	0	65

The table shows that the concentrations of sulfur dioxide have the greatest amplitudes of annual variation, and in 18% of the cases they are more than double the value of the mean annual concentration. Nitrogen dioxide and dust also have large amplitudes, but there is an increase in the number of cases in which the amplitude was negligible (less than half the mean value).   Carbon monoxide has the smallest amplitude, and this is not more than $q_{av.an}$ in 81% of the towns; the highest value was 1.75.

The largest amplitudes of the dust content are observed in towns in which dust storms had taken place, and the mean dust concentrations in the different months were high as a result.   Sulfur dioxide and nitrogen dioxide had high amplitudes in towns which showed a winter pollution maximum, that is, in towns where in winter the increase in the concentration of contaminants due to increased combustion of fuel is attenuated by a high frequency of occurrence of meteorological conditions unfavorable to the dispersion of contaminants (such as Irkutsk).

A meteorological analysis of the data obtained in 1967–1969 showed that there is a correlation between the seasonal variations in the contaminant and the frequency of wind speed between 0 and 1 m/sec. It was noted from observations in more than 30 towns that the increase in weak winds is accompanied by an increase in the mean monthly values of the concentrations of nitrogen oxides, carbon monoxide, and sulfur dioxide. The calculated correlation coefficients between the frequency of weak winds and the mean monthly concentrations of carbon monoxide were 0.47–0.71. The data of /2/ show that the correlation is highest in towns where low discharge sources predominate, and where weak winds are observed over a wide layer and elevated inversions are frequent. An increase in the concentration of sulfur dioxide is frequently noted in months with the largest number of days with fog and haze. This agrees satisfactorily with the theoretical conclusions and physical concepts on the mechanism of propagation of sulfur dioxide in fogs /1/.

Since a high frequency of fogs is accompanied by a high frequency of weak winds and temperature inversions, all these factors together lead to an increase in the air pollution level. Thus, in Krivoi Rog in November 1968, the mean monthly concentration of sulfur dioxide was three times the mean annual concentration, as a result of the increase by a factor of almost 2 of the frequency of elevated inversions and fogs.

Local climatic conditions have a considerable influence on the annual variation. Thus, in Baku an autumn-summer maximum of air pollution by sulfur dioxide is observed, connected with the characteristics of the circulation on the Caspian sea shore. This period is characterized by an increase in the frequency of winds from the sea, with a stably stratified layer flowing from the sea to the land. The frequency of stratification types with elevated inversions is maximum in summer /4/.

To analyze the factors causing seasonal variations in the content of contaminants in the atmosphere, the mean monthly values $\overline{M}$ obtained by averaging these values were calculated for the three contaminants:

$$\overline{M} = \frac{M_{SO_2} + M_{NO_2} + M_{CO}}{3}.$$

This averaging smoothes the fluctuations in the contaminants caused by variations in the method of discharge, and shows more clearly the fluctuations caused by factors operating simultaneously on all

types of contaminants. For each month of 1968, maps of the distribution of $\overline{M}$ (Figure 3) were plotted, on which are clearly distinguished zones with values of $\overline{M}$ above (or below) the mean annual
value, i. e., with higher or lower levels of contaminants. Trajectories of the cyclones and anticyclones at ground level are plotted
on these maps from the daily synoptic bulletins of the meteorological service. It was found that the isolines of mean yearly $\overline{M}$
($\overline{M}=1$) usually lie on the boundary between the zones of movement
of the cyclones and anticyclones. During the cold season zones of
higher contaminant content are formed along the paths of movement
of the little-mobile anticyclones, characterized by weak winds,
temperature inversion at night, and intense turbulent exchange
during the day, absence of directed transfer of the air mass in the
lower layer of the troposphere up to heights of 3—5 km, and absence
of prolonged precipitations. The largest values of $\overline{M}$ are observed
near the regions where the anticyclone continues for a few days,
and the smallest values in the regions of active cyclonic activity.

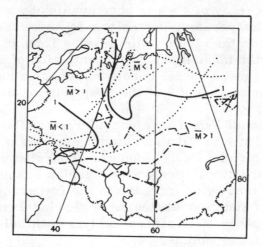

FIGURE 3. Distribution of the relative air pollution.
February 1968.

    During the warm season the distribution of contaminants in the
air is more complex. Usually an increase in the mean monthly
concentrations is observed in regions of movement of cyclones,
with weakly developed barometric field and a pressure at the center
higher than 1,000 millibars, when a weakened turbulent exchange is
observed in a large layer of the troposphere.
    The analysis of the diurnal variation in the concentrations over
the USSR territory was based on observations of the concentrations

of carbon monoxide, nitrogen dioxide and sulfur dioxide in 48 towns
in the cold (October—March) and warm (April—September) seasons
of 1968—1969.

The largest diurnal variations were noted for sulfur dioxide concentrations in towns of the southern regions. The cold season is
characterized by a diurnal variation with clearly marked maxima in
the morning and evening hours (Figure 4a). It is determined by the
fact that at night, as a result of radiation cooling, ground inversions
are usually formed, which are preserved until sunrise; the wind
speed is not high. The dispersion of cold discharges from low
sources at night is weakened by the high stability of the atmosphere.
The beginning of the working day is characterized by an increase
in the discharges of plants and boiler rooms. As a result, the maximum content of contaminants in the atmosphere will be observed in
the morning hours.

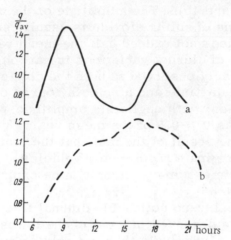

FIGURE 4. Diurnal variation in the concentrations of sulfur dioxide in the cold
(a) and warm (b) periods

After sunrise the disintegration of the inversion starts from its
base. Although the discharges of low sources are preserved, they
now take place in an ever increasing air volume, and therefore the
concentrations of contaminants do not increase. The simultaneous
increase in the wind speed leads to a decrease in the air contamination in the afternoon hours to a minimum.

We should note that under these conditions the high discharge
sources do not contribute greatly to air pollution in the case of a
small mixing layer, since these discharges are above the inversion,

and therefore do not fall to the ground. But when the inversion is destroyed in the morning hours, the air with the contaminants from high sources mixes with the lower layers, which can lead to an additional increase in the concentrations. Sometimes a more significant evening maximum, caused by the increase in stability, may appear.

A daytime maximum in the concentration of sulfur dioxide is observed in the towns of the northern and central parts of the European Territory of the USSR.

This is because in winter the temperature inversion in the northern regions is destroyed later, and the maximum shifts. An analogous diurnal variation is created by an increase in the turbulent exchange in the industrial towns, where the contaminants are discharged from high stacks.

In the warm period the diurnal variation is not expressed so clearly, but a daytime maximum of the concentrations is observed in many towns (Figure 4b). The amplitude of the diurnal variations in the concentrations of sulfur dioxide is usually small, and in most towns (about 70%) does not exceed half the mean value.

Two basic types of diurnal variations in carbon monoxide concentrations are clearly observed in both the cold and warm periods: with a maximum in daytime and a smooth course of the concentrations during the day with negligible amplitude, which in 90% of the cases represents 0.1—0.5 of the mean yearly value. This type of distribution is the result of the fact that the main discharges of carbon monoxide originate from motor vehicles. The existence of a maximum in daytime in many towns can be explained by the traffic peak at these hours.

For nitrogen dioxide, no noticeable diurnal fluctuations in the concentrations were observed in most of the towns studied (amplitude $0.00—0.02 \, \text{mg/m}^3$) during either the cold or warm periods.

We consider below the diurnal variation in air pollution under average conditions. The diurnal course of the concentration can vary considerably as a function of the wind speed and direction, the temperature stratification, the cloudiness, etc. The diurnal cycle of sulfur dioxide concentration for different wind speeds and directions was analyzed from observations in Baku, using the results of a station located at the center of a residential quarter and not directly exposed to the effect of discharge sources.

Figure 5a shows that the highest concentration of sulfur dioxide is observed at 13 hours. The diurnal variations are clearly marked at very low wind speeds (0—2 m/sec) only. Almost no diurnal variation is observed at speeds of 3—10 m/sec. This is obviously because in the residential area an increase in the concentration of

sulfur dioxide takes place only in the case of a weakening in the
wind while an increase in speed is accompanied by a rapid dispersion of the contaminant, and so its content in the air remains
almost constant throughout the day.

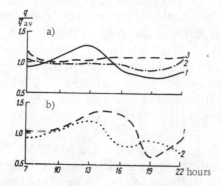

FIGURE 5. Diurnal variation in the concentration of sulfur dioxide as a function of the speed (a) and direction (b)
of the wind:

Wind speed: 1) 0−2 m/sec; 2) 3−5 m/sec;
3) 6−10 m/sec; wind direction: 1) northern; 2) southern.

Figure 5b gives curves of diurnal variations in the concentration
of sulfur dioxide in Baku with northern and southern winds of speed
0−2 m/sec. The figure shows that the air contamination is much
lower with a southern wind than with a northern wind; this is due to
the purifying effect of the Caspian Sea. Besides, with a southern
wind a small increase in concentration is observed at 13 hours only,
while with a northern wind this increase remains throughout the day.

Bibliography

1. Berlyand, M.E., R.I.Onikul, and G.V.Ryabova. K teorii
        atmosfernoi diffuzii v usloviyakh tumana (The Theory of
        Atmospheric Diffusion under Fog Conditions).— Trudy
        GGO, No. 207. 1968.

2. Berlyand, M. E. and R. I. Onikul. Fizicheskie osnovy rascheta rasseivaniya v atmosfere promyshlennykh vybrosov (Physical Principles for Calculating the Dispersion of Industrial Discharges in the Atmosphere). − Trudy GGO, No. 234, pp. 3−27. 1968.

3. Bezuglaya, E. Yu. K opredeleniyu potentsiala zagryazneniya vozdukha (Determination of the Air Pollution Potential). − Trudy GGO, No. 234, pp. 69−79. 1968.

4. Vdovin, B. I. and A. A. Gorchiev. Tipovye profili temperatury v nizhnem kilometrovom sloe atmosfery nad Apsheronskim poluostrovom (Typical Profiles of the Temperature in the Bottom Kilometer Layer of the Atmosphere above the Apsheron Peninsula). − Trudy GGO, No. 238, pp. 195−200. 1969.

5. Son'kin, L. R., E. A. Razbegaeva, and K. M. Terekhova. K voprosu o meteorologicheskoi obuslovlennosti zagryazneniya vozdukha nad gorodami (The Problem of the Meteorological Factors Determining the Pollution of Air above Cities). − Trudy GGO, No. 185. 1966.

6. Tomson, N. M. Sanitarnaya okhrana atmosfernogo vozdukha ot zagryaznenii (Sanitary Protection of Atmospheric Air from Pollution). Leningrad, Medgiz. 1959.

7. Shabad, L. I. and P. P. Dikun. Zagryaznenie atmosfernogo vozdukha kantserogennymi veshchestvami (Contamination of Atmospheric Air by Cancerogenic Substances). Leningrad, Medgiz. 1959.

8. Barrett, C. F. Correlations of Smoke Concentration in Great Britain. − Int. J. Air Wat. Pollut., Vol. 7, pp. 991−993. 1963.

9. Colucci, I. M., C. R. Begeman, and K. Kumler. Lead Concentrations in Detroit, New York and Los Angeles Air. − J. Air. Pollut. Control Ass., Vol. 19, No. 4. 1969.

10. Commins, V. T. and K. E. Waller. Observations from a Ten-Year Study of Pollution at a Site in the City of London. − Atmospheric Environment, Vol. 1. 1967.

11. Gaizago, L. Variation in the SO_2 Content of the Air as a Function of Weather Conditions. − Időjárás, Vol. 2. 1964.

12. Jamoto, M. Air Pollution in Tokyo. A Summary of Aerometric Surveys during 1963−1964. Tokyo, Japan. 1965.

13. Summers, P. W. The Seasonal Weekly and Daily Cycles of Atmospheric Smoke Content in Central Montreal. − J. Air Pollut. Control. Ass., Vol. 16, No. 8. 1968.

THE COULOMETRIC METHOD FOR DETERMINING
THE HALOGEN-CONTAINING COMPOUNDS IN THE AIR

N. Sh. Vol'berg

The problem of determining a large number of harmful substances in the air by automatic instruments can be solved only if
multipurpose gas analyzers are available. We earlier established
the conditions for the coulometric determination of a number of
sulfur-containing compounds (sulfur dioxide, hydrogen sulfide and
carbon disulfide) present simultaneously in the air by a single
instrument /3/. The aim of the present study is to develop conditions for the continuous coulometric determination of free halogens and halogen-containing compounds.

Determination of free halogens

A number of coulometric cells for the determination of chlorine
and other free halogens have been described in the literature /7−9/.
Most of these were designed for the continuous flow or circulation
of the electrolyte. Due to the simplicity and reliability of a cell
without circulation developed for the determination of sulfur
dioxide /1/, we wished to develop a similar design for determining
free halogens as well. The sulfur dioxide cell (the small model)
was adapted as follows. The pyrolusite electrode was replaced by
an electrode of porous silver applied on graphite, and the cell was
filled with different electrolytes. To prevent sorption of the halogen,
gas was introduced into the solution by means of a narrow tube
made of polyfluoroethylene resin.

The silver electrode was prepared by the repeated application of
silver nitrate crystals onto a heated graphite rod, followed by
decomposition of the salt in a furnace at a temperature of 600−800°C.

When air-containing free halogen enters the cell, the following
reactions take place on the electrodes:

$$Cl_2 + 2e \rightarrow 2Cl^- \quad \text{and} \quad Ag - e + Cl^- \rightarrow \underset{\downarrow}{AgCl}.$$

The continuous passage of air through the limited volume of the electrolyte in a cell without circulation may lead to the rapid evaporation of the water and to the crystallization of the dissolved salts on the electrodes.   This causes a drop in the current output. Therefore, considerable attention was paid to the development of the composition of the electrolyte for filling the coulometric cell.

It was considered necessary to introduce into the composition of the electrolyte compounds which prevent dessication and favor the dissolution of free halogens in the aqueous medium.   The organic liquids usually used, such as glycerin, ethylene glycol, etc., are undesirable because of the possibility of a chemical reaction.   We recommend an electrolyte consisting of a 25% solution of lithium chloride, characterized by a high hygroscopicity.

To increase the solubility of halogens in water, iodides of alkaline metals are used, which form weak complexes with the iodine liberated.   However, the introduction of iodides into the composition of the electrolyte is undesirable, because of its adverse effect on the operation of the silver electrode.   Therefore, a suitable organic complexant was sought.

We did not succeed in finding a complexant directly for chlorine, and accordingly decided to use the ability of elementary bromine to form complexes, in particular with acetamide.   To convert the chlorine into equivalent amounts of bromine, a small amount of lithium bromide as introduced into the electrolyte of lithium chloride.   The following composition of the electrolyte was selected from data of experiments to check the retentive capacity of chlorine and bromine: lithium chloride, 25%; lithium bromide, 0.02%; acetamide, 4%.

The solution of lithium chloride with these additives was tested directly in the coulometric cell.   The comparison was based on the current strength of a cell containing different types of electrolytes at a constant concentration of the halogen in the air pumped through the cell.

A diffusion doser was selected for the uniform supply of chlorine and bromine /2/.   When chlorine was batched into the ampule, the doser was filled with a solution of chlorine in carbon tetrachloride; when bromine was batched, it was filled with pure bromine.   A constant supply of iodine vapors ensured the passage of air through a layer of its crystals.   The rate of flow of the iodine was limited by a very thin capillary glass tube.   The concentrations of the different halogens were controlled in such a way that the current generated by the cell was between 30 and 40μA.   The results of the measurements obtained with different electrolyte compositions are given in Table 1.

TABLE 1. Value of the current (μA) as a function of the electrolyte composition

Electro- lyte No.	Electrolyte composition	Chlorine	Bromine	Iodine
1	Lithium chloride — 25%	26.0	33.0	41.0
2	Lithium chloride — 25%, acetamide — 4%	28.0	43.5	42.5
3	Lithium chloride — 25%, acetamide — 25%	29.5	44.5	42.5
4	Lithium chloride — 25%, acetamide — 4%, lithium bromide — 0.02%	30.0	45.0	41.5
5	Potassium bromide — 3.5%, sodium iodide — 0.075%, phosphate buffer pH-7	29.0	44.0	42.5
6	Lithium chloride — 25% lithium bromate — 10%	29.0	41.5	41.0

It is seen from the results that the current efficiency is little dependent on the composition of the electrolyte. The effect of the electrolyte is smallest when determining iodine, and the efficiency is almost the same in all the solutions tested. The reason probably lies in the sorption of iodine on the electrodes, as mentioned in a number of papers. Chlorine gives the highest current efficiency in electrolyte 4, bromine gives roughly the same results in all the electrolytes containing acetamide.

Thus, tests on the cell confirmed the advisability of using an electrolyte of the following composition: lithium chloride, 25%; lithium bromide, 0.02%; acetamide, 4%. The background current of a cell with this electrolyte and an electrode of gold gauze of area 8 cm^3 was 0.3μA, and was reproducible within ± 0.1μA.

The optimum rate of supply of chlorine-containing air was found by special experiments. Gas was supplied at a constant rate from the doser into the air stream pumped through the cell, and the dependence of the magnitude of the generated current on the air speed was determined. It follows from the results that in the range between 30 and 100 ml/min, the current magnitude is independent of the speed, while at higher (or lower) speeds it decreases negligibly.

The current efficiency generated by the coulometric cell according to the electrically active substance supplied to it was estimated by comparison with the results of chemical analysis. The amount of chlorine or bromine emerging from the diffusion doser in the same time was determined in parallel.

The comparative experiments were carried out at relatively large concentrations to ensure an optimum accuracy of the photo-metric measurements: $10-40$ mg/m^3 for chlorine and $2-10$ mg/m^3 for bromine.   The current efficiency measured was found to be near to the theoretical value, and equal to $98 \pm 2\%$ and $94 \pm 3\%$ for $f = 9$ and 5 respectively, and $\alpha = 0.95$ (where f is the number of degrees of freedom and α the order of accuracy).

The results indicate that the cell developed can be used for determining microconcentrations of chlorine without preliminary calibration.

Determination of hydrogen chloride

The coulometric determination of hydrogen chloride and other hydrogen halides is possible after their oxidation to free halogen.

From our study on the reactions of hydrogen chloride with different oxidants we concluded that the reaction with potassium bromate is the most promising, since it is not accompanied by losses of chlorine during the oxidation of metal-containing anions or oxides.

Tychkova and Filyanskaya /5/ studied the properties of a number of oxidants when they were developing a linear coloristic method for determining HCl in air, and also concluded that potassium bro-mate is the most suitable oxidant for hydrogen chloride.

To increase the chemical activity of neutral potassium bromate, we applied this salt in the form of a saturated solution on porous materials with a more or less developed specific surface.   Large-pore carriers were used to avoid chlorine adsorption.   The inert carriers used in gas chromatography are satisfactory: diatomic carrier TND-TS-M, and hydrothermally processed silica gel with a specific surface of 70 mg^2/g — silochrom-3.

Porous polytetrafluoroethylene, which has a very low adsorptive capacity, was also tested.

The oxidants were prepared as follows.   The carriers were boiled with dilute 1:1 HCl, and washed with hot distilled water to a neutral reaction.   They were then annealed for 1 hour at 900°C. The required quantity of powder (fraction 0.2—0.25 mm) was heated in a drying cupboard to 100°C, and uniformly wetted by a hot solution of potassium bromate.   The weight was selected so as to ensure that the finished product would contain 20% of dry salt by weight. The powder was then dried for 2—3 hours at 110—120°, with occasional stirring.

The polytetrafluoroethylene carrier was prepared as follows. The polytetrafluoroethylene powder was poured onto an aluminium plate in a 20-mm thick layer, and placed for 30 minutes inside a muffle furnace heated to 380°.   To avoid local overheating, the plate was placed on porcelain supports.   The baked mass was then taken out, cut into small pieces with scissors, and ground in an electric coffee grinder, and the 0.25—0.5 mm fraction was sifted.   The powder was wetted by an equal volume of acetone, and an equal volume of an aqueous solution of potassium bromate, saturated at 30°, was poured over it.   The potassium bromate precipitated uniformly in the carrier pores during mixing.   A volume of acetone equal to the initial volume was then poured on the preparation, the excess liquid was poured off, and the residue was dried first in the air, and then in a drying cupboard at 60°.

The different oxidants were tested for given concentrations of hydrogen chloride, obtained in the diffusion batcher /2/.

A solution of lithium chloride in hydrochloric acid was used as the source of hydrogen chloride.   Before the air was fed to the doser, it was dried by a TND carrier impregnated by potassium chloride or lithium chloride.

The oxidant to be tested was placed in a vertical glass tube on a pad of quartz wadding.   The rate of supply of the hydrogen chloride-air mixture was 100 ml/min.   The current strength in the coulo-metric cell, which is proportional to the quantity of hydrogen chloride oxidized to free halogen, was measured in the experiments. The results (mean values of several measurements) are given in Table 2.

TABLE 2. Influence of the carrier on the oxidation of hydrogen chloride by potassium bromate (current output in μA)

Carrier	Oxidant volume (ml)								
	0.01	0.025	0.05	0.1	0.2	0.3	0.4	0.5	1.0
Polytetrafluoroethylene ..	—	16	22	23	26	27.4	31	31	—
TND	—	30	30	31	26.5	—	26	—	—
Silochrome-3	28.5	28.5	28.5	29.5	29.5	—	—	29.5	30.5

The results of the measurements indicate the high oxidizing activity of KBrO$_3$ applied on silochrome-3 and TND; more than 90% of the hydrogen chloride (at a concentration of the order of $7\mu g/l$

and a rate of 100 ml/min) is oxidized by a layer whose volume is 0.01 ml. Potassium bromate on polytetrafluoroethylene has a lower oxidizing capacity: 0.4 ml of powder are necessary to achieve complete oxidation.

When TND was used as oxidant, the maximum output of free halogen was observed with a layer volume of 0.1 ml. The current decreased when the amount of oxidant was increased further, apparently because of the interaction of free halogen with the carrier material. This undesirable phenomenon was not observed in tests with the oxidant on the silochrom -3 carrier.

Next, the influence of the air humidity on the completeness of oxidation of HCl was studied. In a series of experiments it was found that the oxidizing activity of the layer drops considerably in dry air: with an oxidant amount of 0.3 ml, the readings in dry air are approximately $\frac{1}{3}$ lower than usual.

Heating the layer also reduced its oxidizing capacity. Thus, at a constant concentration of HCl, increase in the oxidant temperature to 100°C led to a drop in the current in the coulometric cell from 155 to 80 μA, that is, the output of chlorine dropped to approximately one half. When the layer was cooled, its oxidizing capacity was completely restored, which indicates that there is no decomposition. Since different conditions may occur when the instrument is used it seemed very desirable to select an oxidant that is insensitive to temperature variations. The drop in the oxidant activity with increase in temperature and with dry air seems to indicate that a small amount of moisture must be contained in the layer for a successful reaction.

To maintain sufficient moisture, a hygroscopic salt, lithium chloride, was introduced into the oxidant composition; the most favorable results were obtained with 5% of lithium chloride. Since this chemisorbant has the same oxidizing capacity at room temperature as that without LiCl, there was relatively little change even when the substance was heated to 100°C. With an oxidant volume of 0.3 ml, the current output dropped by only 8%, while without lithium chloride it dropped to one half.

Thus, to oxidize HCl at a flow rate of the air to be analyzed of 100 ml/min, we recommend 0.2 ml of silochrome-3, containing 20% of potassium bromate by weight and 5% of lithium chloride.

The completeness of oxidation of hydrogen chloride under these conditions was checked by comparing the results of the determination of the concentration of HCl in the air by a chemical method (by the reaction with mercury thiocyanate) and by the coulometric method.

The gaseous mixtures were obtained dynamically, and contained between 50 and 200 mg/m^3 of HCl. The current output was $92 \pm 3.5\%$ at $f = 14$ and $\alpha = 0.95$.

Determination of chlorinated hydrocarbons

For the coulometric determination of chlorinated hydrocarbons, it was first necessary to produce a burning device of small weight and dimensions, through which the air to be analyzed could be supplied at the rate of 100 ml/min, according to the technical characteristics of the coulometric cell.

To achieve a maximum intensification of the process, we used catalytic burning in the presence of finely powdered platinum on the carrier. Catalysts containing different amounts of platinum on carriers of different structure were tested. A high degree of dispersion of platinum was obtained by reducing its chloride by a solution of ascorbic acid neutralized by sodium carbonate. The catalysts were prepared as follows. The carrier was uniformly impregnated by a solution of platinic chloride in water with a calculated 1 or 5% of platinum, dried in a drying cupboard at 200°C, and boiled for 15 min in a small amount of a 10% solution of ascorbic acid, first neutralized to pH-7 by sodium carbonate.

The catalyst was washed thoroughly by boiling distilled water (to a negative reaction for chlorine ions, and then three more times), dried in a drying cupboard, and annealed for an hour at a temperature of 800°. The catalyst was tested in a vertical quartz tube of 3 mm diameter, heated by an MA-G furnace with a 50 mm long incandescent zone. The furnace temperature was measured, and maintained at the given level by an EPV potentiometer. The catalyst was placed on a wad of quartz wadding introduced into the tube. The lower end of the tube with the catalyst was attached to a capsule containing 0.2 ml of oxidant, connected to the coulometric cell. Air was continually pumped through the system at the rate of 100 ml/min.

We tested the efficiency of the catalyst by burning dichlorethane vapors fed at a constant flow by means of a diffusion doser /2/. In the experiments the current of the coulometric cell was measured.

This current was generated by the entry of the chlorine that was formed by the oxidation of hydrogen chloride, the product of the catalytic burning of dichloroethane. Table 3 gives the values of this current as a percent of the current in the coulometric cell taken as 100%, and obtained during the combustion of dichlorethane in a vacuum quartz tube of 6 mm diameter in a furnace with a 190 mm long incandescent zone, at a temperature of 900° and rate of air flow of 30 ml/min.

It is seen from the results given in the table that at a temperature of 900° all the catalysts tested based on dispersed platinum ensure complete combustion of dichlorethane even at layer volumes of 0.05 ml; at 700° this is ensured by silica gel with a small specific surface but high platinum content of 5%, and by silica gel with a relatively large specific surface of 346 m^2/g. Platinum applied on aluminum oxide has a particularly high oxidizing capacity, ensuring quantitative combustion even at a temperature of 600°. However, the slowness of the desorption of the hydrogen chloride makes it unsatisfactory in practice.

TABLE 3. Comparative results of tests on different combustion catalysts

No.	Carrier	Specific surface (m^2/g)	Platinum content (%)	Volume of catalyst (ml)	Dichlorethane (%) oxidized at temperature, °C			
					900	800	700	600
1	Vacuum tube	—	—	—	88	38.3	0	—
2	—	—	99.9	Pt gauze 4.5 cm^2	95	67		
3	Silochrome-1 ...	26	5	0.05	100	100	98	
4	Silochrome-3 ...	70	1	0.05	100	95	57	
5	INZ-600	≈ 40	5	0.05	100	89		
6	Chromosorb	1–2	1	0.05	100	95	60	
7	Silica gel	346	1	0.025		91	73	
				0.05	100	100	97	85
				0.1		100	99	92
8	Aluminum oxide .	not known	5	0.05	100	100	100	100

Note. Conditions of the experiments: tube diameter 2.6 mm, rate of air supply 0.1 l/min, concentration of dichlorethane 10 mg/m^3.

Thus, from the experiments, as a catalyst for the combustion of dichlorethane we can recommend large-pore silica gel with platinum applied on it (specific surface 300–400 m^2/g), containing not less than 1% of platinum, in a layer of 0.1 ml, and at a temperature of 700°.

After the determination of the optimum conditions for the catalytic combustion of dichlorethane, oxidation of the hydrogen chloride formed, and coulometric determination of free halogen, the method was tested for given concentrations of dichlorethane.

To obtain a continuous flow of accurately known low concentrations of the vapors of a volatile liquid, a diffusion batcher with optical measurement of its rate of flow was developed /4/.

By statistical processing of the results of the measurements it was found that in the range of concentrations between 2 and 37 mg/m^3, the current efficiency is 95% of the theoretical value, with a probable standard deviation from the arithmetic mean (for $\alpha = 0.95$) of ± 2.2% at $f = 17$. When very low vapor concentrations were determined of the order of 0.4 mg/m^3, the current efficiency dropped to 80%.

We next tested the possibility of determining other chlorinated hydrocarbons by the method described. It was found that methylene chloride gives a current efficiency of 82% of the theoretical value, and chloroform 92%. Carbon tetrachloride is not oxidized under the given conditions.

Since chlorine and hydrogen chloride interfere with the coulometric determination of chlorinated hydrocarbons, we tested a number of substances for absorbing them. The best results were obtained with a mixture of potassium carbonate and sodium thiosulfate applied on the inert porous carrier TND (sferochrom-1) and colored by phenolphthalein. The chlorine and hydrogen chloride are satisfactorily absorbed, but not the vapors of volatile chlorinated hydrocarbons. The sorbent becomes decolorized during the process.

Thus, these studies led to the development of a coulometric method for determining different chlorine-containing substances in the air. The method can be considered, in principle, as a general one for determining chlorine in organic compounds (not perchlorinated), but the observed deviations of the current efficiency from the theoretical value show that an experimental test in each particular case is necessary.

From the results a prototype of an automatic gas analyzer was designed for free halogens, hydrogen chloride, and chlorinated hydrocarbons. In the determination of chlorine, the air to be analyzed is fed directly to the coulometric cell, and for hydrogen chloride it is first passed through a tube with an oxidant, silochrom-3, containing 20% of potassium bromate and 5% of lithium chloride. The chlorinated hydrocarbons are determined after combustion on platinized large-pore silica gel at a temperature of 700—800°C, and oxidation by potassium bromate. To remove chlorine and hydrogen chloride in the determination of chlorinated hydrocarbons, we used a capsule with a mixture of potassium carbonate and sodium thiosulfate applied on a porous carrier.

For a rate of flow of the air to be analyzed of 0.1 l/min, and a secondary instrument with a scale of $0-5\mu A$, the gas analyzer described has a sensitivity of the order of tenths of a milligram per cubic meter of air and an inertia of 10—15 min.

Summing up, we note that the use of preliminary reactions with solid sorbents and catalysts is a very promising means for producing multipurpose continuous-action gas analyzers. The advantage of such instruments over the other known multipurpose gas analyzers, such as photocolorimetric ones, lies in the possibility of a rapid transition from the measurement of the content of one compound to the measurement of another one by simply turning the multichannel gas stopcock.

Bibliography

1. Al'perin, V. Z. et al. Usovershenstvovannaya gal'vanicheskaya kulono-polyarograficheskaya yacheika dlya izmereniya malykh kontsentratsii dvuokisi sery (Improvement of the Galvanic Coulomb—Polarographic Cell for Measuring Low Concentrations of Sulfur Dioxide). — In: Avtomatizatsiya khimicheskikh proizvodstv, No. 1, pp. 76—82. Moskva. 1968.
2. Vol'berg, N. Sh. Ustroistva dlya dinamicheskoi podachi malykh kolichestv gazov i parov v potok vozdukha (Devices for the Dynamic Supply of Small Amounts of Gases and Vapors to the Air Flow). — In: Metody opredeleniya vrednykh veshchestv v vozdukhe, pp. 130—134. Leningrad. 1968.
3. Vol'berg, N. Sh. Kulonometricheskii metod opredeleniya serusoderzhashchikh soedinenii v vozdukhe (Coulometric Method for Determining Sulfur-Containing Compounds in the Air). — Trudy GGO, No. 238, pp. 107—114. 1969.
4. Vol'berg, N. Sh. Dozator parov letuchikh zhidkostei (Doser of Vapors of Volatile Liquids). — Trudy GGO, No. 238, pp. 229—231. 1969.
5. Tychkova, I. E. and E. D. Filyanskaya. Lineino-koloristicheskii metod opredeleniya khloristogo vodoroda (Linear-Coloristic Method for Determining Hydrogen Chloride). — Trudy VNIIOT, p. 280. 1967.
6. Dubois, L., A. Zdrojewski, and Monkman. Analysis of Carbon Monoxide in Urban Air at the ppm Level, and the Normal Carbon Monoxide Value. — J. Air Pollut. Control Ass., Vol. 16, No. 3, pp. 135—140. 1966.

7. Ersepke, Z. and J. Baranek.   Analysátor chlóru v ovzduší. —
        Chem. Průmysl, Vol. 16, No. 8, pp. 496—497. 1966.
8. Souček, J.   Coulometrické stanovení chlóru v plynech. — Chem.
        Průmysl, Vol. 13, pp. 470—471. 1963.
9. Waclawik, J. and S. Waszak.   Ciagłe oznaczanie małych
        ilości chloru w gazach. — Chemia analit., Vol. 12,
        pp. 877—883. 1967.

DETERMINATION OF HYDROGEN FLUORIDE IN ATMOSPHERIC AIR

T. A. Kuz'mina, N. Sh. Vol'berg

The methods used for determining HF in atmospheric air and the air of industrial enterprises are based mainly on the ability of fluorine to form stable colorless compounds with several polyvalent cations, and thus prevent the formation of colored complexes of these cations with the corresponding organic reagents /2/. Both titrimetric and photometric variants of these methods are used.

Thus, the method of back-titration by a solution of thorium nitrate with alizarin red as indicator is widely used /3, 9/. The advantage of the method consists in the possibility of determining fluorine ions over a wide range of concentrations, from $1 \mu g$ to 50 mg in the sample. However, at low concentrations of fluorine ions the method is rather subjective, due to the difficulty of accurately determining the color transition at the equivalence point.

Of the photometric methods, the most widely used is that based on the decomposition of the titanium chromotropic complex /1,4,5/. The advantages of this method are the relative simplicity of the analyses and the satisfactory specificity. The minimum determinable amount is $1.5 \mu g$ per 5 ml sample.

Photometric methods based on the use of compounds of thorium, zirconium, lanthanum and cerium with different organic reagents (arsenaso I, pyrocatechol violet, eriochrome cyanine, xylenol orange, thoron, etc.) have a high sensitivity /8/.

Finkel'shtein used a thorium-thoron complex for determining fluorine compounds in the air /7/. With his method $2-40 \mu g$ of fluorine can be determined in a sample of 25 ml. The presence of aluminium in 2 to 5 times the amount of fluorine does not interfere with the determination. The influence of oxidants is eliminated by introducing $1-2$ ml of a 5% solution of hydroxylamine hydrochloride into the solution.

For serial determinations of hydrogen fluoride in atmospheric air, it was found advisable to replace radioactive thorium by non-toxic and easily available zirconium. According to published data, the complex of zirconium with xylenol orange has the highest sensitivity for determining fluorine in different substances

/8, 10, 11/. We used this complex in the development of a method
for determining HF in atmospheric air.

The pH is of great importance in reactions of this type, and
accordingly the first problem is to find the optimum acidity for
determining fluorine. We therefore measured the optical density
of solutions containing constant amounts of sodium fluoride
($C = 2\,\mu g/5$ ml and $C = 4\,\mu g/5$ ml) at different concentrations of
hydrochloric acid. The optical density was measured by a photo-
electrocolorimeter FEK-56 with light filter No. 6, with a transmission maximum at 540 mμ.

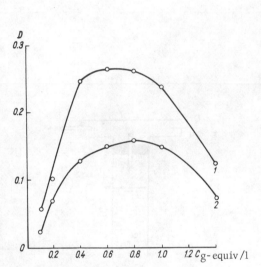

FIGURE 1. Optical density of solutions of
the complex of zirconium with xylenol
orange as a function of the concentration
of hydrochloric acid:

1) with a fluorine ion content of 2 μg per
5 ml sample; 2) with a fluorine ion content
of 4 μg per 5 ml sample.

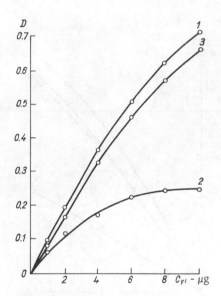

FIGURE 2. Optical density of the reagent
solution as a function of the concentration of
fluorine ions (C_F, μg/5 ml):

1) in absence of SO_4^{2-}; 2) in presence of
162,000 μg/ml SO_4^{2-} (in the form of 3.5 N
H_2SO_4); 3) in presence of 880 μg/ml
SO_4^{2-} (in the form of a solution of Na_2SO_4).

The relationship between the optical density and the acidity of
the medium is shown in Figure 1. The curve shows that in the
range of concentrations of hydrochloric acid between 0.4 and 1 N, the
sensitivity of the determination is highest, and is little dependent on
variations in the acidity of the solutions. A concentration of 0.6 N
is optimum. To obtain such an acidity, the zirconyl nitrate must be
dissolved in 3.5 N hydrochloric acid.

The interference of SO_4^{2-} and Al^{3+}, which usually accompany hydrogen fluoride, was the subject of a special study.   To solutions containing different amounts of fluorine ions were added known amounts of SO_4^{2-} (in the form of H_2SO_4 and Na_2SO_4) and Al^{3+} (in the form of $Al(NO_3)_3$).   The results (Figure 2) indicate that SO_4^{2-} inter- feres at relatively high concentrations only not usually found in the air. Al^{3+} has a considerable effect.   Thus, even at a concentration of aluminium ions of $2\mu g/ml$, there is a certain deviation from the curve plotted without aluminium, while at a concentration of $20\mu g/ml$, 40% of the given amount of fluorine ions is determined (Figure 3).

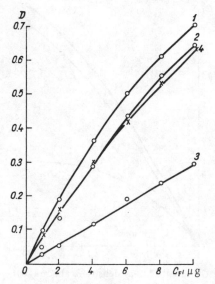

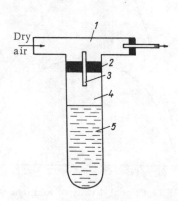

FIGURE 3. Optical density of the reagent solution as a function of the concentration of fluorine ions (C_F, $\mu g/5\,ml$):

1) in absence of Al^{3-}; 2) in presence of $2\,\mu g/ml$ Al^{3+}; 3) in presence of $20\,\mu g/ml\,Al^{3+}$; 4) in presence of $120\,\mu g/ml\,Al^{3+}$ with pre- liminary heating.

FIGURE 4. Batcher of hydrogen fluoride:

1) polyethylene T-junction; 2) polyethy- lene stopper; 3) Teflon capillary tube inserted in the stopper; 4) polyethylene vessel; 5) 1:1 mixture of concentrated sulfuric and hydrofluoric acids.

By comparing the instability constants of fluorine complexes /6/, we note that the complexes of fluorine with zirconium are more stable than the complexes with aluminium.   Therefore, zirconium displaces aluminium from its combinations with fluorine.   The influence of aluminium salts can be explained by the slowness of the displacement process.

The results of the experiments indicate that heating on a water bath for 30 min appreciably reduces the influence of aluminium on the results of the determination of fluorine ions. In this case it is possible to determine F^- in the presence of a 12-fold amount of aluminium.

When the method was developed, particular attention was paid to questions of sampling and sample storage. To find the admissible rate of feeding air through the system of two serially connected absorbing instruments, air containing hydrogen fluoride was passed at different speeds. A constant concentration of HF was obtained by means of the batcher shown in Figure 4. The amount of fluorine ions was varied between 1.5 and $14\mu g$ in the sample. It was found by experiment that at flow rates of up to $5\,l/min$ the amount of hydrogen fluoride determined in the second absorbing instrument does not exceed 5% of the given amount, and for $3\,l/min$, 3%. Thus, to determine fluorine ions in atmospheric air, sampling must be performed at a flow rate of $2-3\,l/min$. This ensures almost complete capture in the first absorbing instrument.

Since the efficiency of absorbing instruments with a porous plate and of Richter absorbing instruments was found to be roughly the same, it was found to be more efficient to use Richter absorbers, which have a smaller surface of contact with the solution.

A possible interaction between hydrogen fluoride and the glass of the absorbing instruments was taken into account, and the admissible time of storage of the samples was determined (Table 1).

TABLE 1. Variation in the concentration of F^- during storage in glass absorbing instruments

Given F^- (μg)	0.25	0.50	1.00	2.00	4.00	5.00	6.00	8.00	10.00
Found F^-, after a day (μg)	0.28	0.60	0.89	1.99	4.00	4.87	5.50	7.80	9.60

The results show that the concentrations of fluorine ions decrease in the course of a day, although negligibly. It is therefore desirable to analyze the samples on the day of sampling.

The following procedure based on the studies carried out can be recommended. The air to be investigated is passed through a Richter absorbing instrument, filled with 5 ml of doubly distilled water, at the rate of $3\,l/min$. To free the air from aerosols of compounds of aluminium and sulfates, and also to capture solid fluorides, a cartridge with filter AFA-V-10 is placed in front of the absorbing instrument. In the laboratory, to wash off the F^- adsorbed on the glass of the inlet tube, water is poured into the absorbing instrument up to the 6 ml mark. The solution is poured out through the same tube. In the analysis, 1.6 ml of zirconyl nitrate solution (80 mg $ZrO(NO_3)_2 \cdot 2H_2O$) are added to 5 ml of the sample solution, and dissolved in 1 l of titrated 3.5 N hydrochloric acid. The solution is thoroughly stirred, and held for 30 minutes. Simultaneously, five blank solutions are prepared by adding 1.6 ml of zirconyl nitrate solution to 5 ml of water. After 30 min, to all the samples (including the blanks) 2.6 ml of a 0.02% solution of xylenol orange are added (a 0.2% solution is prepared and stored for 3—5 days; on the day of the analysis it is diluted to a 0.02% solution) and the solutions are vigorously stirred to obtain an average blank sample, the five are mixed together. After 3—5 minutes, the optical density of the solutions is measured at $\lambda = 540$ mμ with reference to the average blank sample. The measurements are carried out in a cell with $l = 20$ mm.

TABLE 2. Reproducibility of the analysis results ($n = 18$ and $\alpha = 0.95$)

Content of F$^-$ (μg)	0.125	0.25	0.50	1.00	2.00	4.00	6.00	8.00	10.00
Probable root-mean-square deviation from the arithmetic mean (μg)	0.03	0.07	0.04	0.06	0.10	0.15	0.10	0.10	0.10
Relative error of the determination (%)	24	28	8	6	5	3.7	1.7	1.25	1.5

The concentration of hydrogen fluoride in the sample is found from a calibration curve plotted from measurements of the optical density of a series of standards. The standard solutions are prepared in 100 ml flasks in such a way that each 5 ml of solution contains 0.125, 0.25, 0.5, 1.0, 2.0, 4.0, 6.0, 8.0, and 10.0 μg, respectively, of fluorine ions. The calibration curve is plotted from the mean values calculated from the results of the measurement of four-five standards. The error in the determination of the different concentrations of fluorine ions in the statistical processing of the results of the analysis of 18 series of standards, carried out at different times by different authors, is given in Table 2, calculated with an accuracy of $\alpha = 0.95$.

The sensitivity of the method is 0.002 mg/m^3, for a sampling rate of 3 l/min and a sampling time of 20 min. The error in the determination of the maximum permissible concentration in atmospheric air (0.02 mg/m^3) is ± 6%.

22048

Bibliography

1. Babko, N. K. and P. V. Khodulina. Tsvetnaya reaktsiya na
   ion ftora s titan-khromotropovym reaktivom (Color
   Reaction for Fluorine Ions with a Titanium Chromotropic
   Reagent). — Zhurnal Analiticheskoi Khimii, No. 5. 1952.
2. Kiseleva, E. K. Analiz ftorsoderzhashchikh soedinenii
   (Analysis of Fluorine-Containing Compounds). Moskva,
   "Khimiya." 1966.
3. Metodika kontrolya za sostoyaniem vozdushnoi sredy v elektro-
   liznykh tsekhakh alyuminievykh zavodov (Method for
   Controlling the State of the Air in Electrolysis Shops of
   Aluminum Plants). Sverdlovsk. 1960.
4. Panin, K. P. Metod opredeleniya malykh kolichestv ftora v
   atmosfernom vozdukhe s primeneniem titankhromo-
   tropovogo reaktiva (Method for Determining Small Amounts
   of Fluorine in Atmospheric Air with a Titanium Chromotropic Reagent). — Gigiena i Sanitariya, No. 9. 1959.
5. Panin, K. P. Ob opredelenii ftora v atmosfernom vozdukhe
   (Determination of Fluorine in Atmospheric Air). — Gigiena i
   Sanitariya, No. 12. 1967.
6. Spravochnik khimika (Handbook for Chemists). Vol. 3, p. 39.
   Moskva, "Khimiya." 1968.
7. Finkel'shtein, D. N. Izbiratel'nyi metod kolorimetrichesko-
   go opredeleniya floristykh soedinenii v vozdukhe (The
   Method for Sampling in the Colorimetric Determination of
   Fluorine Compounds in the Air). — Gigiena i Sanitariya,
   No. 9. 1969.
8. Tušl, J. Fotometrické metody stanovení fluoru. — Chemické
   Listy, Vol. 61, No. 10. 1967.
9. Matuszak, M. P. and D. R. Brown. Thorium Nitrate Titration of Fluoride with Special Reference to Determining
   Fluorine and Sulfur in Hydrocarbons. — Ind. Engng. Chem.
   Analyt. Edn., Vol. 17, No. 12. 1945.
10. Novák, J. and H. Areno. — Silikáty, Vol. 8, No. 1. 1964.
11. Novák, R. Patent ČSSR, class 42e, 3/53, No. 105355.

SOME RESULTS OF THE DETERMINATION OF THE CONCENTRATION OF SULFUR DIOXIDE IN THE FREE ATMOSPHERE

O. P. Petrenchuk, V. M. Drozdova

Gaseous combinations of sulfur contribute considerably to the pollution of atmospheric air. The most widespread contaminant in the atmosphere is sulfur dioxide, which is released mainly in the combustion of fuel. However, the studies of a number of authors show that sulfur compounds, both gaseous and in the form of aerosols, are present everywhere, not only in the troposphere, but in the stratosphere as well.

Recently, interest in the study of sulfur compounds in the atmosphere has increased considerably with the discovery in the stratosphere, at a height of 20—25 km, of the so-called sulfatic layer, made of sulfate particles /8—10/. There are different opinions on its origin. C. Junge and a number of other authors /8, 9/ explain the existence of this layer by the oxidization of the gaseous compounds of sulfur, mainly sulfur dioxide and hydrogen sulfide, and the photochemical reactions in which they participate. It is considered that the sulfur compounds mentioned above have a tropospheric origin. Martell /10/ indicates that the boundaries of this layer are not clear, and assumes that the layer is principally due to the coagulation of Aitken nuclei, formed at great altitudes mainly of sulfates. As there are few experimental results on the vertical distribution of sulfur dioxide in the atmosphere, we cannot estimate the relative part played by its oxidation or by the coagulation of Aitken nuclei in the process of formation of the sulfatic layer in the lower stratosphere.

Sulfates are always contained in cloud water, fogs and precipitations, and they are frequently the predominant contaminant /1, 4, 5, 9/. From the results of experimental studies /7, 8/ it has been assumed that one of the important sources of sulfates in atmospheric waters is the oxidation of sulfur dioxide in the droplets in the presence of catalysts. Therefore, when studying the formation of the chemical composition of precipitations, it is important to know the content of sulfur dioxide in the free atmosphere. However, such data are at present very scarce /6/.

Therefore, in March—April 1969, we carried out aircraft measurements of the content of sulfur dioxide in the free atmosphere near
one of the cities in the Northwestern European Territory of the
USSR.   An IL-14 airplane was used, and flights were performed
under different meteorological conditions.   During the given period,
there were 15 flights, and 15 vertical profiles of the content of sulfur dioxide in the atmosphere were obtained up to a height of about
2,500 m.

The concentration of sulfur dioxide was measured by the West
and Gaeke method, one of the most sensitive of the conventional
methods, and giving the concentration of sulfur dioxide in the sample
to within $0.1 \mu g$ /11, 2/.   The analytical method of West and Gaeke
has a great advantage over the other methods, as the gas is trapped
as such, and the presence of sulfate aerosols does not affect it.
This method is based on the fixation of sulfur dioxide when air is
passed through a 0.1 N solution of mercuric chloride and sodium
chloride, $Na_2(HgCl_4)$, and the formation of the stable nonvolatile
disulfitomercurate ion.

Air samples were taken by the rotating device PRU-4.   The air
to be studied was aspirated at the rate of 5 l/min for 15—20 min
through a U-shaped absorber with a porous partition, filled with
5 ml of absorbant solution.   Special studies /6/ showed that the
maximum possible error in the sampling caused by fluctuations in
the rate of air flow during sampling, by losses of the absorbant
solution due to splashing and evaporation, by the adsorption of traces
of gas in the connecting pieces, and by the influence temperature is
about ±10%.

After the sample was taken, care was taken during the laboratory
analysis that the volume of the absorbant solution, which might
decrease as a result of splashing, be maintained constant and equal
to the initial volume.   This was achieved by adding an additional
amount of absorbant solution before the analysis.   The analysis
was carried out by the colorimetric method with pararosaniline
hydrochloride and formaldehyde.   The final product of the reaction
in this method is the purple-colored fairly stable pararosaniline-
sulfonic acid.   The calibration curve was plotted by using freshly
prepared working solutions, containing 20 and $2 \mu g$/ml of SO_2,
respectively.   The content of sulfur dioxide was determined in the
initial solution from which the working solutions were prepared by
weighing the sample and by titrating by 0.01 N iodine.   Benzene was
used as indicator in the titration, as it is more sensitive than starch
/3/.   All the solutions were prepared with freshly obtained doubly
distilled water.

Before the colorimetric measurement, 0.5 ml of pararosaniline solution and 0.5 ml of a 2% solution of formaldehyde, were added to the test tubes with samples, and the test tubes were carefully covered.   After 25 min a red-violet color developed, which persisted for about 30 min.   The optical density of the solutions was measured on a FEK-M photocolorimeter in cells with $l = 10$ mm, and a green light filter.

All the determinations of the content of sulfur dioxide were carried out not more than 1—3 days after the sampling, although, as preliminary laboratory tests showed, the samples are stable for up to 8 days.   At the end of this period, a noticeable decrease in the concentration of sulfur dioxide is observed, and on the 12th day of storage it may be 25—40% of the initial value.

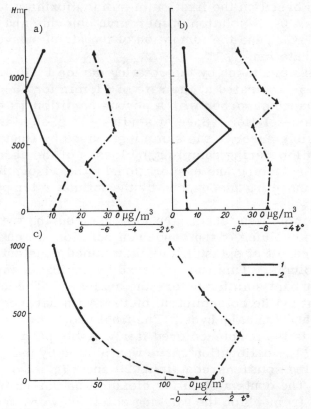

FIGURE 1. Variation in the concentration of sulfur dioxide and the temperature of the air with height under anticyclone conditions in March 1969:

a) 24 March; b) 26 March; c) 19 March; 1) sulfur dioxide; 2) temperature.

In the period of flights beginning on 15 March, all the European Territory of the USSR was covered by a wide low-gradient high-pressure zone, which remained till the end of the month. The weather in the Northwestern European Territory of the USSR was characterized by very weak winds and cloudlessness. Industrial haze existed above the town for many days. From 31 March, 1969, the barometric conditions changed. A wide deep cyclone was formed above Scandinavia. The Northwestern European Territory of the USSR, including the region of the flight was in the zone of clearly marked direct transfer from the Southwest.

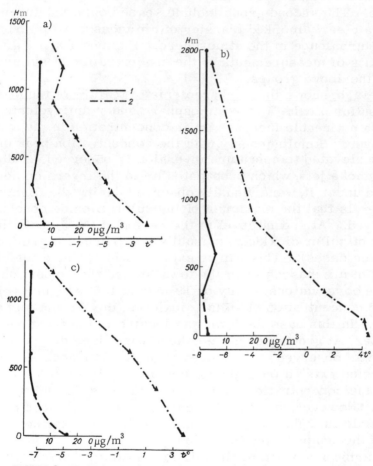

FIGURE 2. Variation in the concentration of sulfur dioxide and the temperature of the air with height under conditions of clearly marked transfer of air masses in April 1969:

a) 3 April; b) 14 April; 10 hours; c) 14 April, 20 hours; 1) sulfur dioxide; 2) temperature.

On the days of the flights, a first sample was taken before the flight at the airport at ground level; the next samples were taken by the plane at heights of 300, 600, 900, 1,200 m, and in some cases at higher heights as well (up to 2,600 m). At these levels the plane circled the town at a radius of 15—20 km. Each sampling took about 20 minutes. At the end of the flights the samples were sent to the chemical laboratory for analysis.

The results of measurements on the concentration of sulfur dioxide in the atmosphere can be classified into two groups, according to the observed synoptic situation. The first group includes all the cases of measurements carried out under anticyclonic conditions, characterized by stability, weak winds, and temperature inversions. The second group includes cases obtained under conditions of a clearly marked transfer of air masses, with strongly developed turbulence in the atmosphere. Figures 1 and 2 give some results of measurements of the concentration of sulfur dioxide typical of the above groups.

Figure 1a, b, shows that a characteristic feature of the distribution of sulfur dioxide in the atmosphere under anticyclonic conditions is a noticeable increase in its concentration in the subinversion layer. Simultaneously with the concentration peak of sulfur dioxide, an elevated temperature inversion is observed. Industrial haze and smoke jets, which propagated up to the inversion height and spread under it, were visually observed during the flights. It is at these levels that the maximum of the sulfur dioxide concentration was observed. At ground level in the region of the airport the concentration of sulfur dioxide was usually low (about $10 \mu g/m^3$, or even less in some cases). This can be explained by the absence of movement of air masses from the town toward the airport on the days of the observations. Only on 19 March 1969 (Figure 1c) was the ground concentration of sulfur dioxide in the airport higher than $120 \mu g/m^3$. In this case the town was located in the front part of the anticyclone. At ground level a northern wind (i. e., directed from the town) was observed with a speed of up to 10 m/sec. A temperature inversion existed from the ground up to a level of 330 m. At this level the concentration of sulfur dioxide was fairly high (about $50 \mu g/m^3$). However, since in this case the ground concentration of sulfur dioxide is high, there is no high maximum of the concentration in spite of the temperature inversion.

The distribution of sulfur dioxide in the atmosphere when there was clearly marked transfer of air masses is totally different. Figure 3 shows typical profiles of the sulfur dioxide concentration and the temperature in the atmosphere under these conditions. On 3 April 1969 (see Figure 2a) the town was located in the region of

direct transfer of air masses from the Northwest.   The wind speed
at ground level was 8—10 m/sec.   On 14 April 1969 (see Figure 2b,c)
the town was located at the rear of the cyclone; a southwestern wind
with a speed of 13—18 m/sec was observed at ground level.

The high wind speed at ground level and at a high level, and the
turbulent exchange developed lead to equalization of the concentrations of sulfur dioxide in the atmosphere.   In this case, as seen
from the figures, the concentration of sulfur dioxide at ground level
is lower than under anticyclonic conditions.   It varies little with
height, that is, an almost homogeneous distribution of sulfur dioxide
is established in the atmosphere.   Even at a height of 2,600 m
(see Figure 2b) the concentration of sulfur dioxide is not equal to
zero, and differs little from that at lower levels.

Thus, the studies prove the permanent presence of sulfur dioxide
in the atmosphere, and the appreciable variation in its content with
height in the lower layers of the troposphere, depending on the
meteorological conditions.

Bibliography

1. Drozdova, V. M. et al.   Khimicheskii sostav atmosfernykh
      osadkov na Evropeiskoi territorii SSSR (Chemical Com-
      position of Atmospheric Precipitations over the European
      Territory of the USSR).   Leningrad, Gidrometeoizdat.
      1964.
2. Zasukhin, E. N. and V. P. Cherepakhova.   Fotokolori-
      metricheskoe opredelenie sernistogo gaza v atmosfernom
      vozdukhe (Photocolorimetric Determination of Sulfur
      Dioxide in Atmospheric Air). — Trudy GGO, No. 185. 1965.
3. Lavrinenko, R. F.   O soderzhanii sery v atmosfernykh
      osadkakh (On the Sulfur Content in the Atmospheric Preci-
      pitations). — Trudy GGO, No. 207. 1968.
4. Petrenchuk, O. P.   Khimicheskii sostav oblachnoi vody v
      raionakh Zapadnoi Sibiri (Chemical Composition of Cloud
      Water in the Regions of Western Siberia). — Trudy GGO,
      No. 234, pp. 130—136. 1968.
5. Petrenchuk, O. P. and V. M. Drozdova.   Khimicheskii
      sostav oblachnoi vody v raionakh promyshlennykh gorodov
      pri raznykh meteorologicheskikh usloviyakh (Chemical
      Composition of Cloud Water in Industrial Regions under
      Different Meteorological Conditions). — Trudy GGO, No. 238,
      pp. 201—209. 1969.

6. Georgii, H.W. and D.Jost.   Untersuchung über die Ver-
teilung von Spurgasen in der freien Atmosphäre. — Pure
and Appl. Geophys., Vol. 59. 1964.
7. Junge, C.E. and T.G.Ryan.   Study of the SO_2 Oxidation in
Solution in its Role in Atmospheric Chemistry. — Q. Jl. R.
Met. Soc., Vol. 84, No. 359. 1958.
8. Junge, C.E., C.W. Chagnon, and J.E.Manson.   A World-
Wide Stratospheric Aerosol Layer. — Science, No. 133.
1961.
9. Junge, C.E.   Air Chemistry and Radioactivity. New York,
London, Academic Press. 1963.
10. Martell, E.A.   The Size Distribution and Interaction of
Radioactive and Natural Aerosols in the Stratosphere. —
Tellus, Vol. 18, Nos. 2—3. 1966.
11. West, P.W. and G.C.Gaeke.   Fixation of Sulfur Dioxide as
Disulfitomercurate and Subsequent Colorimetric Esti-
mation. — Analyt. Chem., Vol. 28, No. 12. 1956.

THE LAWS OF VARIATION IN pH IN ATMOSPHERIC PRECIPITATIONS

R. F. Lavrinenko

The pH value of atmospheric precipitations is one of the basic physicochemical characteristics. It enables us to estimate the type of components in the precipitations, and to form some idea of their origin. The pH value of atmospheric precipitations varies over wide limits, between 4 and 7. Values lower than 4 or higher than 7 are sometimes found.

The aim of the present study was to find which components in the precipitations determine the pH value, and how the proximity of an industrial town affects the variation in pH. We examined the characteristics of precipitations at different pH values. Systematic analyses of precipitations are carried out at the Main Geophysical Observatory, but in January—February 1970, the author carefully took snow samples in the settlement of Voeikovo (15 km from Leningrad).

In their analysis, attention was paid mainly to determining the ions SO_3^{2-}, SO_4^{2-}, HCO_3^-, Ca^{2+}, Mg^{2+}, NH_4^+, the pH, and the electrical conductivity $\varkappa$. The pH was determined by a glass electrode on the pH-meter LPU-01. The methods used for determining the chemical composition of the samples of atmospheric precipitations are described in /2/. The ion SO_3^{2-} was determined immediately after the fall of the precipitations by the method described in /3/, and at the same moment pH and $\varkappa$ were measured. Whenever possible the determination of SO_3^{2-}, pH, and $\varkappa$, was repeated in the samples after several days. The concentration of SO_3^{2-} decreased gradually to 0 in all cases, the pH and $\varkappa$ varied negligibly, although it would appear that the pH should decrease with oxidation of SO_3^{2-} to SO_4^{2-}. The small number of samples did not make it possible to investigate the expected decrease in pH in a large number of cases. The ion SO_4^{2-} was determined only after several days, when $SO_3^{2-} = 0$.

Of the natural agents affecting the pH value in the atmospheric precipitations, the most important is CO_2. At the equilibrium point for atmospheric precipitation, the pH value can be considered as equal to 5.5, which corresponds to the mean content of CO_2 in the air at 20°C /4/. According to Alekin /1/, the pH of dissolved carbon dioxide is equal to 5.6.

The above equilibrium value of the pH can be considered as
sufficiently stable, and the observed variations in CO_2 in the
atmosphere affect it little.   Thus, the expected 25% increase in the
concentration of CO_2 in the atmosphere of the earth by the end of
the century must lead to a 25% increase in the concentration of
hydrogen ions in natural waters, or to a decrease in the pH of only
0.1 (the concentration of hydrogen ions varies almost linearly with
the concentration of free CO_2).

TABLE 1. Frequency of number of samples with different pH values in the precipitations at
Voeikovo

Year	pH					
	4.0–5.5	5.5–5.8	> 5.8	4.0–5.5	5.5–5.8	> 5.8
	Summer period			Winter period		
1962	84	15	1	16	13	6
1963	49	11	2	17	3	3
1964	49	15	5	16	9	–
1965	65	17	14	26	10	11
1966	74	7	10	19	12	7
1967	52	7	13	15	7	9
1968	53	14	10	6	8	14
1969	55	15	15	7	4	19
Mean,%	74	15	11	47	26	27

Table 1 shows what fraction of the precipitations in Voeikovo
corresponds to the natural state of the atmosphere according to the
chemical composition and the pH value of 5.5—5.8.   This type of
precipitations (we shall call it equilibrium or neutral) is characterized by low electrical conductivities of the order of
$(7-20) \cdot 10^{-6} ohms^{-1} cm^{-1}$ and by a correspondingly low mineralization.   Table 2a gives data on the chemical composition of the
precipitations, collected by the author in the winter of 1970 in
Voeikovo.   These can serve as an example of the given type of
precipitations.

It is seen from Table 1 that most of the samples belong to the
acid type, with pH varying from 4.0 to the equilibrium point 5.5.
The acid character of the precipitations is mainly due to the
products of fuel combustion, especially sulfur dioxide, SO_2.   The
annual world discharge of SO_2 is 80 million tons, of which between
50 and 60 million tons are derived from coal combustion and
10 million tons from crude oil /6/.

TABLE 2. Chemical composition of samples of precipitations collected in Voeikovo in January–February 1970

Date of sampling	Content of ions (mg/l)										pH	$\varkappa \cdot 10^6$ (ohm^{-1}·cm^{-1}) at $t = 20°C$
	SO_3^{2-}	SO_4^{2-}	Cl^-	NO_3^-	HCO_3^-	Na^+	K^+	Mg^{2+}	Ca^{2+}	NH_4^+		
a)												
6 Jan.	0.62	2.00	0.43	0.00	1.22	0.67	0.00	0.46	0.00	0.04	5.9	7.7
14 "	–	1.85	–	0.08	0.31	0.20	–	0.00	0.22	0.25	5.6	6.5
14–15 "	–	4.70	1.31	0.06	–	2.80	–	–	0.03	0.29	6.0	16.0
16 "	–	2.40	0.85	0.00	0.85	0.20	0.10	0.10	0.10	0.75	5.9	8.4
22 "	–	1.00	0.57	0.47	1.83	0.16	0.20	0.91	1.22	0.14	5.6	17.7
14 Feb.	0.08	2.85	–	0.73	–	–	–	–	–	0.26	5.5	14.8
b)												
4 Feb.	2.42	6.00	–	–	–	–	–	–	–	0.76	4.5	32.5
8 "	3.98	6.60	–	–	–	–	–	–	–	0.80	5.4	26.0
9–10 "	0.85	1.58	–	–	–	–	–	0.27	0.48	0.38	5.0	17.3
10–11 "	0.20	1.50	0.21	0.90	0.00	0.19	0.10	0.18	0.27	0.41	4.7	18.8
14–15 "	0.96	3.00	0.39	1.12	0.00	0.18	0.20	0.11	0.44	0.08	4.7	20.8
c)												
15–16 Jan.	–	23.00	–	–	–	1.30	–	1.26	3.20	–	5.6	57.5
4–5 Feb.	–	18.90	1.38	1.00	0.00	–	–	2.19	3.00	2.40	5.4	66.0
6–8 "	–	15.40	–	–	–	–	–	2.92	3.10	0.73	6.0	53.4
13 "	12.40	8.62	–	–	–	2.70	0.75	–	–	1.12	5.5	27.7
2 "	–	30.50	1.63	–	–	–	–	–	8.00	0.82	6.7	89.0
4 "	–	11.80	–	0.76	1.83	1.10	0.38	1.85	3.00	1.00	6.1	41.5
12–13 "	11.50	11.50	–	0.58	1.32	0.50	0.27	0.92	3.30	1.04	6.1	35.6
14 "	32.60	22.50	–	–	–	–	–	0.64	6.65	–	7.4	51.8
15–16 "	77.40	31.00	–	–	–	3.70	0.90	1.65	11.30	–	6.6	91.5

Quantitative evaluations of the rate of oxidation of SO_2 to SO_4^{2-} with time in freshly fallen precipitations are given in /3/. Megaw and Cox /5/ studied the absorption of SO_2 by growing and already existing droplets, and found that the coefficient of the rate of absorption drops rapidly with increase in the concentration of SO_2; however, the presence of ammonia increases the rate of absorption of SO_2 by droplets, and the rate of absorption of gaseous SO_2 by a factor of approximately 4. A similar effect of ammonia was studied experimentally by Khenvol and Mezon /6/.

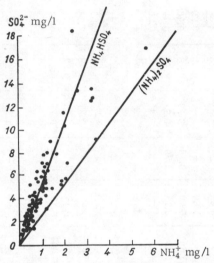

FIGURE 1. Relationship between the concentrations of NH_4^+ and SO_4^{2-} in samples at pH = 4.0–5.5 (according to results of the analysis of samples collected in Voeikovo and Leningrad in 1969).

When we studied the relationship between the concentrations of the NH_4^+ and SO_4^{2-} ions in the precipitations, we observed that in acid samples with low electrical conductivities the concentrations of these ions correspond to the acid salt NH_4HSO_4. Figure 1 shows that the ions NH_4^+ and SO_4^{2-} combine to give the neutral salt $(NH_4)_2SO_4$ in individual cases only. The curves in Figure 1 confirm the conclusions of /5, 6/.

The chemical composition of the acid type of precipitations is illustrated by Table 2b. Precipitations of this type, as seen from Table 1, are characteristic of the industrial region in the summer period.

When coal is burnt, alkaline solid ash particles are formed besides the acid products.   In winter they appreciably alter the character of the precipitations.   In spite of the high concentrations of SO_3^{2-} (see Table 2c), the pH of precipitations contaminated by ash increases considerably.   During solution

$$Ca(Mg)CO_3 + H_2SO_4 = Ca(Mg)SO_4 + H_2O + CO_2 \qquad (1)$$

and neutralization

$$Ca(Mg)(HCO_3)_2 + H_2SO_4 = Ca(Mg)SO_4 + 2H_2O + 2CO_2 \qquad (2)$$

with a generally high mineralization and electrical conductivity, the pH is near to the equilibrium value.

We should note that in Table 1 the equilibrium samples in the winter period also include samples contaminated by acid and alkaline components.   Their neutralization leads to the establishment of a pH value near to the equilibrium value (see Table 2c: 15—16 January, 4—5, 6—8 February).

If solid particles, which create an alkaline medium on solution, are the main contaminant, the pH increases to 7 or more.   Such precipitations are characterized by a high concentration of the ions Ca^{2+}, Mg^{2+}, HCO_3^{1-}, and by a correspondingly high electrical conductivity.

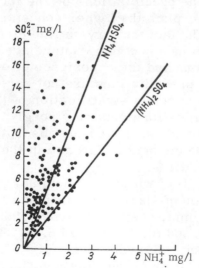

FIGURE 2. Relationship between the concentrations of NH_4^+ and SO_4^{2-} in samples at pH > 5.5.

The graphical representation of the relationship between the ions SO_4^{2-} and NH_4^+ in precipitations with a pH higher than 5.5 (Figure 2) does not show a clear-cut relationship, but here also we see that a large number of points are grouped about the straight lines NH_4HSO_4 and $(NH_4)_2SO_4$. The fact that some of the points are concentrated to the left of the line corresponding to NH_4HSO_4 is probably due to the binding of some SO_4^{2-} to Ca^{2+} and Mg^{2+} ions, according to equations (1) and (2).

Table 2c shows that the concentrations of the SO_3^{2-} ions is greater than that of SO_4^{2-} ions. In /3/ we noted that the measured value of the SO_4^{2-} concentration was lower than the calculated. The concentration of the SO_4^{2-} ions was determined immediately after the thawing of the snow, and the temperature of such a solution is equal to $5-10°C$. With further increase in the temperature (up to approximately $20°C$), the equilibrium of the reaction $SO_2 + H_2O \rightleftharpoons H_2SO_3$ is shifted to the left. The excess of ions SO_3^{2-} over SO_4^{2-} shows that not only an oxidation of SO_3^{2-} to SO_4^{2-} occurs, but also liberation of SO_2.

We should note that the analytical determination of the concentration of ions, especially by the colorimetric method (SO_3^{2-}, SO_4^{2-}, NH_4^+), is complicated in precipitations contaminated by solid contaminants by the formation of stable colloidal suspensions, and the results may be too high.

Many authors /2, 7/ point to a possible dependence of the contaminants contained in the precipitations on the form of their fall as rain or snow. They explain the higher concentrations of contaminants in winter by the fact that crystalline elements fall more slowly than droplets, and have a wider capture area. Our experience in collecting precipitations and the experience of others indicate that in winter also pure precipitations are not rare (see Table 2a). However, in winter, during stable stratification of the atmosphere, the samples of the precipitations may be strongly contaminated by the precipitation of dust and ashes, and the absorption of gaseous contaminants (SO_2) at the moment of fall and in the intervals between falls of precipitations.

A comparison of the pH of winter and summer precipitations indicates that the fraction of alkaline components is higher in winter. The average summer (April—October 1969) value for Voeikovo was pH = 5.38, and for Leningrad 5.28. The average winter (November—March 1969) value was pH = 5.69 for Voeikovo, and 6.61 for Leningrad. In Dresden (Eastern Germany) the average pH value for the period between 1958 and 1962 was 4.5 /4/; the average value of 1964 was 4.6, the average value for the winter period (October-March) 4.41, for the summer period (April—September) 4.49.   These values indicate the high degree of contamination of the atmosphere by acid components in Eastern Germany.

In the Scandinavian network the pH values lie near the equilibrium value of 5.5 /4/, and we can therefore assume that the state of the atmosphere in this region is near to the natural state.

Thus, the following conclusions can be drawn from an analysis of a sufficiently large number of data on the chemical composition of samples of atmospheric precipitations.

The equilibrium pH value corresponds to slightly mineralized precipitations, i. e., to a pure atmosphere.

In acid samples of atmospheric precipitations, the predominant ions in most cases are NH_4^+ and SO_4^{2-}, at concentrations corresponding to the acid salt NH_4HSO_4 or to the neutral salt $(NH_4)_2SO_4$.

The alkaline reaction of the precipitations is caused by the ions Ca^{2+}, Mg^{2+} and HCO_3^-, formed during the leaching of ashes and soil.

If acid and alkaline components are present simultaneously in the atmosphere, the reaction of precipitations may be weakly acid, neutral or alkaline, depending on the ratio between the soluble parts of these components.

Bibliography

1. A l e k i n, O. A.   Osnovy gidrokhimii (Fundamentals of Hydrochemistry). Leningrad, Gidrometeoizdat. 1970.
2. D r o z d o v a, V. M. et al.   Khimicheskii sostav atmosfernykh osadkov na Evropeiskoi territorii SSSR (Chemical Composition of Atmospheric Precipitations in the European Territory of the USSR). Edited by E. S. Selezneva, Leningrad, Gidrometeoizdat. 1964.
3. L a v r i n e n k o, R. F.   O soderzhanii sery v atmosfernykh osadkakh (The Content of Sulfur in Atmospheric Precipitations).— Trudy GGO, No. 207. 1968.
4. M r o s e, C.   Measurements of pH, and Chemical Analyses of Rain, Snow and Fog Water. — Tellus, Vol. 18, Nos. 2—3. 1966.
5. M e g a w, W. J. and L. C. C o x.   A Comparison of the Uptake of Sulfur Dioxide in Growing and Existing Droplets.— Proc. of the 7th Int. Conf. on Condensation and Ice Nuclei. Academia. Prague. 1969.

6. Pierrard, J. M.   Environmental Appraisal. — Particulate Matter,
       Oxides of Sulfur and Sulfuric Acid. — J. Air Pollut. Control
       Ass., Vol. 19, No. 9, Sept. 1969.
7. Petrenchuk, O. P. and E. S. Selezneva.   Izmenenie kontsen-
       tratsii osnovnykh khimicheskikh primesei v osadkakh v
       zavisimosti ot meteorologicheskikh uslovii (Variation
       in the Concentrations of the Main Chemical Contaminants
       in the Precipitations Depending on the Meteorological
       Conditions). — Trudy GGO, No. 134. 1962.

ASSESSMENT OF THE TURBIDITY OF THE ATMOSPHERE FROM ACTINOMETRIC DATA

E. N. Rusina

A knowledge of the overall characteristics of the field of concentrations is necessary in addition to the local characteristics to
estimate the degree of contamination of the atmosphere. Since a
reliable determination of these integral magnitudes requires a dense
measuring mesh, the development of methods for the direct determination of the averaged characteristics is of great importance.
Instruments of the type of the Volz solar photometer are used
abroad /5/. In the Soviet Union it is simplest to determine the
averaged characteristics from actinometric observations, which
give an idea of the general background of the atmospheric contamination over the territory of the Soviet Union and of the degree of
contamination of the different points.

Two parameters are most frequently used in actinometry for the
quantitative evaluation of atmospheric turbidity: the transmission
coefficient P and the turbidity factor T. P is defined as the ratio
of the final intensity of the solar radiation I to the initial intensity
(at the boundary of the atmosphere) I_0 for an optical mass $m=1$.

$$P = \frac{I}{I_0}.$$ (1)

T is the ratio of the logarithms of the transmission coefficient
P and P_i of the real and ideal atmosphere, respectively

$$T = \frac{\ln P}{\ln P_i}.$$ (2)

It is not very convenient to use the transmission coefficient as
a characteristic of the content of aerosols in the atmosphere, since
it increases greatly with increase in the mass of the atmosphere m
as a result of the selectivity of the molecular diffusion. The turbidity factor is less dependent on m, since it indicates the degree
that the transparency of the real atmosphere differs from that of the
ideal atmosphere.

It is known that T can be represented as the ratio of the optical densities Θ and Θ_i of the real and the ideal atmosphere

$$T = \frac{\Theta}{\Theta_i}. \tag{3}$$

In its turn

$$\Theta = \int\limits_{0}^{\infty} a_i \rho_i \, dh + \int\limits_{0}^{\infty} a_w \rho_w \, dh + \int\limits_{0}^{\infty} a_d \rho_d \, dh, \tag{4}$$

where a_i is the molecular diffusion coefficient, a_w and a_d are the attenuation coefficients of the radiation by water vapor and dust, ρ_i and ρ_w are the densities of air and water vapor, ρ_d is the concentration of dust.

Thus, the turbidity factor helps us to isolate the components of the overall attenuation of the radiation caused by the Rayleigh scattering, and also by the moist and residual (aerosol) turbidity. We therefore selected the turbidity factor to estimate the background of atmospheric contamination. This is easily determined from observations on the direct radiation at a given optical mass, or can be calculated by the Sivkov method /2/ from the values of the direct solar radiation reduced to the mean distance between Earth and Sun and to the same height of the Sun.

Although actinometric characteristics have been used since the twenties to study transparency of the atmosphere, relatively little attention has been paid to an assessment of the background of atmospheric contamination over large territories.

The most complete study on the distribution of the transmission coefficient of the atmosphere over the territory of the Soviet Union, using the radiation characteristics of many years, was carried out by Pivovarova /4/. She gives maps of the distribution of the transmission coefficient for the four most characteristic months of the year. It was found that the yearly cycle of the transparency is determined mainly by variation in the absolute humidity, that is, it has a maximum in winter and a minimum in summer. Regions with highest and lowest transparency were determined, and the distribution of the value of the aerosol attenuation over the territory was studied.

This study showed that the atmosphere is most contaminated over the European Territory of the USSR, Middle Asia, and the large industrial regions. The influence of city conditions on the transparency of the atmosphere was also studied in the paper. In contrast to /4/, we consider here the background characteristics of

the distribution of atmospheric turbidity over the territory of the
USSR in each month. Particular attention is paid to the application
of the actinometric method for estimating the degree of pollution
of the atmosphere in cities.

To obtain the general background of the atmospheric turbidity
above the territory of the Soviet Union, we plotted maps of the
intensity of the direct solar radiation falling on a horizontal surface
(S') under conditions of clear sky in the main actinometric periods
(the maps were based on the data of handbooks on the climate of the
USSR, part 1, for each of the five periods). We took into account
that in all the periods of the observations the isolines of the intensity show a well-expressed latitude dependence without appreciable inflections and closed regions (except for the locations of
lower intensity of solar radiation round large industrial towns), and
read from the maps the mean latitude values of the solar radiation
intensity in each period. The values of the turbidity factor in all
the months of the year were calculated by Sivkov's method using
these data. The values were averaged, and the mean values of the
turbidity factor for the month (T) were obtained for different latitudes between 45 and 75°. Table 1 gives values for 5° latitude
intervals.

TABLE 1. Distribution of the mean values of the turbidity factor in different latitudes during the year

Latitude (deg)	Jan.	Feb.	Mar.	Apr.	May	June	July	Aug.	Sep.	Oct.	Nov.	Dec.
45	2.24	2.38	2.30	2.67	2.85	2.91	3.38	3.01	2.60	2.66	1.87	1.95
50	1.49	2.54	2.04	2.76	2.85	3.15	3.14	2.72	2.68	2.26	2.06	1.74
55		2.18	2.06	2.79	2.88	3.00	3.05	2.91	2.76	2.41	2.00	
60		2.24	2.03	2.49	2.65	2.77	2.82	2.62	2.41	2.14	1.80	
65			2.13	2.64	2.37	2.36	2.65	2.37	1.98	1.18		
70			2.73	2.29	2.26	2.30	2.45	2.25	1.96	2.25		

Table 1 shows that over the territory of the USSR the turbidity
of the atmosphere decreases gradually with increase in the latitude.
This is easily explained by the general geographical and meteorological conditions (the existence of steppe and desert surfaces and
large areas of arable land, the stronger winds, etc.). However,
against the general background of decrease in the turbidity factor
from South to North from the beginning of spring until the middle
of autumn, a certain increase in the turbidity is noted in the region

of 50—55°N, which is apparently due to the concentration of industries in this belt, and the contamination of the atmosphere by industrial discharges.

A rearrangement of the optical properties of the atmosphere takes place at the beginning of spring (March) and in the middle or even the end of autumn.

While the turbidity factor increases in general in spring, due to the melting of snows and the increase in the condensation turbidity, in middle latitudes a certain decrease in T is observed in March, apparently caused by the increased exchange between the upper and lower layers of the atmosphere. Such a mechanism was noted in the yearly cycle of the radioactive contamination of the atmosphere /3/, and the transfer of the contaminant across the tropopause /1/. The drop in the turbidity in autumn may be due to the change in the character of the precipitations (the increase in the length of covering precipitations, the number of days with precipitations, and the washing out of the aerosols). In the summer period, in spite of the increase in the sum of precipitations in summer months, the short duration of these precipitations and their shower-like character prevents them from purifying the atmosphere, especially because the strong convective currents cause the turbidity to propagate in this period to a much higher altitude.

Table 1 enables us to follow the yearly cycle of the atmospheric turbidity at each latitude. However, since it is impossible to determine the value of T in winter months in northern latitudes, we can form an idea of the variation in the turbidity factor during the year at medium latitudes only, where the yearly cycle of T is simple, with a maximum in June—July and a minimum in December—January.

From the values of the intensities S' in the summer months, during which there are five periods of observations, it was possible to follow the variation in the turbidity factor during the day at different latitudes (Table 2).

TABLE 2. Diurnal cycle of the turbidity factor in summer (June—August)

Latitude (deg)	Time (hours, min)					
	6.30	9.30	12.30	15.30	18.30	Mean
40	2.62	3.08	3.31	3.36	–	–
45	2.59	2.84	3.18	3.17	–	–
50	2.63	2.93	3.06	3.02	2.25	2.78
55	2.83	2.85	3.05	3.07	2.67	2.89
60	2.22	2.63	2.63	2.88	3.05	2.68
65	2.14	2.17	2.45	2.59	2.74	2.42
70	1.92	2.21	2.34	2.39	2.40	2.25

Table 2 shows that the turbidity is maximum at midday.

We have now obtained an idea of the general background of atmospheric turbidity above the territory of the USSR, and shall consider the possibility of using the actinometric method for estimating town pollution.   Unfortunately, parallel actinometric observations in stations inside and outside the cities are very scarce.   We were able to select three such groups:

1) Leningrad — Voeikovo — Lar'yanskaya — Nikolaevskoe,
2) Moscow University — Sobakino,
3) Irkutsk — Khomutovo.

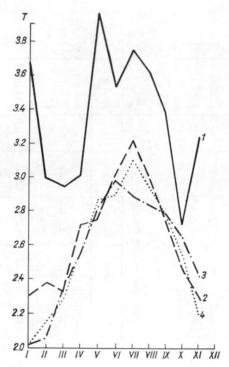

FIGURE 1.   Yearly cycle of the turbidity factor:

a) Leningrad; 2) Voeikovo; 3) Lar'yanskaya; 4) Nikolaevskoe.

From observations on direct solar radiation, we calculated the mean (over many years) values of the turbidity factor for each month of the year for these groups of stations.   Figure 1 represents the yearly cycle of the turbidity factor, averaged over all the actinometric periods for the Leningrad group of stations.   Figure 1

TABLE 3. Mean monthly value of the turbidity factor for Leningrad and its suburbs

	Jan.	Feb.	Mar.	Apr.	May	June	July	Aug.	Sep.	Oct.	Nov.	Dec.	Mean
Leningrad	3.67	2.99	2.94	3.01	3.95	3.53	3.75	3.61	3.38	2.72	3.24	–	3.34
Suburbs	2.11	2.19	2.32	2.59	2.81	2.96	3.07	2.98	2.77	2.56	2.31	–	2.61
Difference	1.56	0.80	0.62	0.42	1.14	0.57	0.68	0.63	0.61	0.16	0.93	–	0.73
Difference (%) ...	74.0	36.5	26.8	16.2	40.5	19.4	22.1	21.1	22.0	6.3	40.0	–	28.0

TABLE 4. Yearly turbidity factors at the stations of Irkutsk and Khomutovo

	Jan.	Feb.	Mar.	Apr.	May	June	July	Aug.	Sep.	Oct.	Nov.	Dec.	Mean
Irkutsk	2.26	2.45	2.65	2.77	3.14	2.98	3.00	2.87	2.64	2.35	2.22	2.59	2.65
Khomutovo	1.90	2.55	2.54	2.70	2.89	2.85	2.89	2.69	2.37	2.08	2.02	2.14	2.44
Difference	0.36	0.20	0.11	0.07	0.25	0.13	0.11	0.18	0.27	0.27	0.22	0.45	0.21
Difference (%) ...	19	8.9	4.3	2.6	8.7	4.5	3.8	6.7	11.4	13	11	21	8.6

shows that the yearly cycle and the values of the turbidity factor in Leningrad and in suburbs differ sharply. In Leningrad two maxima are observed in May (3.95) and January (3.66), and two minima, in March (2.94) and in October (2.72). The suburbs display a simple cycle, with a maximum in July and a minimum in January. The summer maximum of the turbidity factor in Leningrad and its suburbs is due to the increased aerosol turbidity. The winter maximum is observed within the city only; it is particularly large in Leningrad, and weaker in Moscow, probably due to the sea climate of Leningrad.

The air turbidity in the suburbs is much lower than in the city. In the suburb of Voeikovo, which is more heavily polluted than the other suburbs due to its proximity to the city, the value of the turbidity factor is, on the average during the year, 25% lower than in Leningrad. Table 3 gives the mean monthly values of T for Leningrad and its suburbs (the results of the Voeikovo, Lar'yanskaya and Nikolaevskoe stations were averaged). It follows that the average yearly turbidity of the atmosphere in the city is 28% higher than in the suburbs, the difference being greatest in January (74%), when a winter maximum of T is observed in the city, and in May and November (40%) due to the presence of secondary maxima of the turbidity factor, probably due to an increase in the condensation turbidity at this period of the year.

A characteristic feature of Leningrad, as well as of the other towns, is the high purification of the atmosphere from contaminants in spring and autumn. As a result of this purification, in March and October the transparency of the atmosphere in the city approaches that in the suburbs. This very important fact deserves special attention.

Table 1 shows that in autumn, and in particular in spring, a certain self-purification of the atmosphere in general is characteristic of middle latitudes (see, for example, the data for March or September at a latitude of 60°). However, the variation in the turbidity factor, on an average, for all latitudes is not very significant, and at the three stations surrounding Leningrad is not noted. Under urban conditions, when the atmosphere is highly polluted, the mechanism of natural self-purification of the atmosphere plays an important part.

The regularities in the variation in atmospheric turbidity found for the Leningrad group of stations also apply to the pair of stations Moscow-Sobakino (Figure 2). On an average, during the year the atmosphere in Moscow is polluted 14% more strongly than in the suburbs. The yearly cycle of T is similar to the Leningrad cycle, with the difference that the winter maximum is much less marked, since due to the continental climate the condensation turbidity does not play as important a role as in Leningrad. In autumn and spring the differences in the degree of contamination of the atmosphere.in Moscow and its suburbs are negligible, and as in the preceding case, this is due to the self-purification of the city air in these periods.

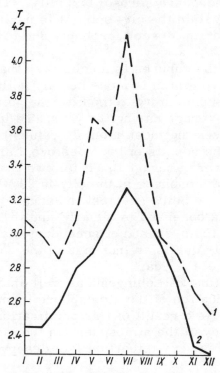

FIGURE 2. Yearly cycle of the turbidity factor at midday:

1) Moscow; 2) Sobakino.

The features observed above hold also for the pair of stations Irkutsk-Khomutovo. Table 4 shows a comparison of the values of the turbidity factor in Irkutsk and Khomutovo in different months of the year. On an average for the year, the turbidity in the city is 8.6% higher than in the suburbs. The largest difference in the values of the turbidity factor of the city and suburbs is observed in winter (19—20%), the smallest difference in early spring and summer. The following conclusions can be drawn.

1. From the climatic norms of direct solar radiation under a clear sky it is possible to establish the turbidity background of the

atmosphere over the territory of the Soviet Union. This can be used as a standard for estimating the degree of contamination of different regions.

2. The following can be deduced from an analysis of this background:

a) the turbidity decreases in general from South to North, but a zone of increased turbidity is clearly seen in middle latitudes, due probably to industrial contamination;

b) the yearly cycle of atmospheric turbidity in southern and moderate latitudes is simple, with a maximum in July, and minimum in December—January. In northern latitudes it has two maxima and two minima;

c) in middle latitudes a decrease in T is observed in spring, probably due to an increased exchange between the upper and lower atmosphere.

3. The degree of contamination of the atmosphere in cities can be estimated from parallel actinomctric observations in city and suburb stations.

The analysis of three pairs of such stations showed that:

a) in large cities (such as Leningrad, etc.) the mean annual values of the turbidity factor are 13—28% higher than in the suburbs; in winter this difference may reach 70%;

b) in spring and autumn, in middle latitude cities there is a peculiar self-purification of the air, and as a result the values of the turbidity factor approach those observed in the suburbs;

c) the annual cycle of atmospheric turbidity in cities is totally different from that in the suburbs. Besides the maximum at the end of the spring or beginning of summer, observed in the city and in its suburbs, a second maximum is noted in winter in large cities, especially in cities with a marine climate.

Bibliography

1. Makhta, L. Perenos primesi v stratosfere i ee perenos cherez tropopauzu (Transfer of a Contaminant in the Stratosphere and its Transfer across the Tropopause). — In: Atmosfernaya diffuziya i zagryaznenie vozdukha. Moskva, Izdatel'stvo inostrannoi literatury. 1962.
2. Metodicheskie ukazaniya po opredeleniyu kharakteristik prozrachnosti atmosfery dlya aktinometricheskikh otdelov (grupp) gidrometeorologicheskikh observatorii (Methodical Instructions for Determining the Transmission

Characteristics of the Atmosphere for the Actinometric Sections of Hydrometeorological Observatories). — GGO. 1965.

3. Marly, U. G.   Znacheniya meteorologicheskikh protsessov v radioaktivnom zagryaznenii atmosfery (Importance of Meteorological Processes in the Radioactive Pollution of the Atmosphere). — In: Atmosfernaya diffuziya i zagryaznenie vozdukha. Moskva, Izdatel'stvo inostrannoi literatury. 1962.

4. Pivovarova, Z. I.   Raspredelenie koeffitsienta prozrachnosti atmosfery (dlya integral'nogo potoka) po territorii SSSR (Distribution of the Transmission Coefficient of the Atmosphere (for Integral Flow) over the Territory of the USSR). — Trudy GGO, No. 213. 1968.

5. Volz, F. E.   Some Results of Turbidity Networks. — Tellus, Vol. 21, No. 5. 1969.

AUTOMATIC GAS ANALYZER AND SOME RESULTS OF THE RECORDING OF CARBON MONOXIDE IN ATMOSPHERIC AIR

E. A. Pevzner, A. S. Zaitsev

The organization of a system of operational control of the pollution of the atmosphere in cities is very dependent on the development of automatic gas analyzers for continuous recording of the concentrations of harmful components. With an automatic instrument we can not only greatly increase the volume of information on the degree of air contamination, but also eliminate errors during the sampling of air, the transport of samples to chemical laboratories, and manual chemical analysis. The first attempt to use a gas analyzer to record sulfur dioxide /2, 3/ showed the high efficiency of the application of automatic instruments, and led to a number of important results.

An automatic equipment for recording the concentrations of carbon monoxide in atmospheric air is important because CO is one of the main components of atmospheric contamination. Carbon monoxide is highly toxic, and remains for a long time in the atmosphere.

The main sources of the emission of carbon monoxide are traffic (mainly motor cars), industrial enterprises (casting industries, oil purification, cellulose plants, etc.) and various combustion processes (combustion of solid wastes, forest and industrial fires, burning of coal wastes, etc.). According to an analysis of the emissions in the USA in 1968 /6/, of a total carbon monoxide emission of 100 million tons, 64% come from the first group, 10% from the second, and 25% from the third.

Existing chemical methods for determining the concentrations of carbon monoxide in atmospheric air /1/ have a number of disadvantages, the chief being the time- and labor-consuming analyses, and the nonspecificity of the determination.

One of the most promising principles for developing automatic gas analyzers for carbon monoxide is the optical-acoustic. However, present Soviet gas analyzers of this type /4/ measure the concentration of carbon monoxide over the range of $0-600$ mg/m^3 within $\pm10\%$.

If we consider that the single highest permissible concentration of CO in atmospheric air is $3 \, \text{mg/m}^3$, and the mean daily concentration is $1 \, \text{mg/m}^3$, the necessity of developing a highly sensitive instrument is obvious.

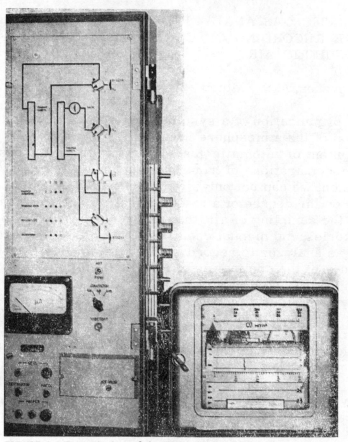

FIGURE 1. General view of the GMK-3 gas analyzer

This paper describes the automatic optical-acoustic gas analyzer GMK-3 developed at the Smolensk branch of the Scientific-Research Institute for Technical Instruments, with the participation of the Main Geophysical Observatory for recording microconcentrations of carbon monoxide in atmospheric air, and gives some results of the analysis of the recording of concentrations of CO in a large industrial city.

The operation of the gas analyzer GMK-3 (Figure 1) is based on measurement of the absorption of infrared radiation by carbon

monoxide.   The degree of absorption of the radiation depends on the
concentration of carbon monoxide in the gaseous mixture.   The gas
analyzer has three measurement ranges: 0—40, 0—80, and
0—400 mg/m^3, which corresponds to scales in volume percentages:
0—0.003, 0—0.006, and 0—0.03, with a basic error of the measurement
of ±5%.   The same working cell is used with all three scales.

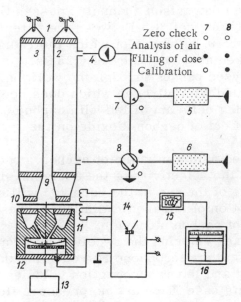

FIGURE 2. Diagram of the gas analyzer:

1) emitter; 2) working cell; 3) comparison cell;
4) diaphragm pump; 5 and 6) filters; 7) batching
valve; 8) three-way valve; 9) filter cells;
10) shutter; 11) receiver; 12) membrane of
electrostatic microphone; 13) motor of shutter;
14) electric measuring circuit; 15) indicating
instrument; 16) recorder.

In the Hartman & Braun (Germany), Choriba (Japan), and Beck-
man (USA) gas analyzers, in the Soviet analyzer GIP-10, and in
other instruments /4/, the problem of increasing the sensitivity
while preserving a sufficiently high signal-to-noise ratio is solved
by lengthening the working cell to 1 m, or by raising the pressure
in the working cell.   This leads to a nonlinear scale, complication
of the instrument, and a number of operational disadvantages.

In the GMK-3 gas analyzer, the length of the working cell is
500 mm, and the gas to be analyzed is at atmospheric pressure.   By

a number of changes in design, we have been able to reduce the magnitude of the residual adjustment signal and the noise. A differential optical system with direct reading is used in the gas analyzer (Figure 2).

The fluxes of infrared radiation are reflected from the metallic mirrors of emitters 1, and enter two optical channels. In the working channel the radiation flux passes through working cell 2 and filtering cell 9; in the comparison channel it passes through comparison cell 3 and filtering cell 9. Between the filtering cells and the receiver is shutter 10, by means of which the radiation fluxes are successively broken at a frequency of 6.25 Hz.

The gas to be analyzed passes through the working cell, and the comparison cell is filled with nitrogen which does not absorb the radiation. The filtering cells are filled with carbon dioxide, and serve to reduce the effect of carbon dioxide on the readings of the gas analyzer.

Receiver 11 is filled with a mixture of carbon monoxide and argon. This ensures the selectivity of the analysis, since the temperature and pressure of the gas in the receiver will change only as a result of the absorption of infrared radiation, corresponding to the absorption spectrum of carbon monoxide.

If the radiation fluxes entering the receiver are equal, the membrane of electrostatic microphone 12 receiving the sum of pressures in the right and left ray-reception volumes will be stationary, since in this case there are no pressure fluctuations in the premembrane volume.

If the radiation flux entering the right ray-reception chamber is reduced as a result of the absorption in the working cell of part of the radiation corresponding to the absorption spectrum of carbon monoxide, in the premembrane volume of the receiver a variable pressure component appears, with a magnitude depending on the degree of absorption of radiation in the working cell. The amplitude of oscillations of the electrostatic microphone membrane is determined by the concentration of carbon monoxide in the gas mixture to be analyzed.

The oscillations of the electrostatic microphone membrane are transformed by the microphone transformer into an alternating voltage, which is amplified by the measuring amplifier, rectified by a synchronous detector, and then transformed into a unified DC output signal between 0 and 5 mA.

A system of alternating comparison of the radiation fluxes of the working and comparison channels is used in the gas analyzer. This makes it possible to develop a receiver with high symmetry of the optical-acoustical channels. The position of the shutter axis located directly in front of the receiver can be controlled smoothly with reference to the optical channels. It is thus possible to obtain a sufficiently low level of the alignment signal. To reduce the vibration noises of the microphone, the membrane is made of a thin polyfluoroethylene film with an aluminium layer sprayed on it.

To increase the sensitivity, the receiver is designed with a smaller passive premembrane volume, and with a somewhat greater power of the receivers than the existing gas analyzers. The sensitivity was also increased by using a less rigid polyfluoroethylene membrane.

A closed emitter was designed to increase the stability of the radiation fluxes. The voltage of the emitter channel is stabilized by a semiconducting stabilizer.

There are no easy methods for the preparation and testing of calibrating gaseous mixtures, and this seriously hampers the attainment of increased accuracy of analyzers of gas microconcentrations. A built-in batcher was developed for the GMK-3 gas analyzer, and it was thus possible to obtain calibrating mixtures inside the instrument with the aid of a single cylinder of carbon monoxide. To obtain calibrating mixtures, nitrogen or air purified by means of chemical absorbants from carbon monoxide and water vapor is blown through the working cell system, which is then closed by the valve 8. The nul point of the instrument is then adjusted. The batching valve 7 is switched to the blowing position, and carbon monoxide is blown through.

By rotating the batching valve, the gas enclosed in the valve is introduced into the working cell chamber. The diaphragm pump 4 is connected, the gas is mixed, and the first calibrating mixture is obtained. By repeating the operation of blowing through the batching valve and introducing a new portion of carbon monoxide in the working cell, we can obtain the second and following calibrating mixtures.

The electric measuring circuit is made up completely of semiconductors. The condenser microphone of the receiver is included in the circuit of the autogenerator convertor. Due to the high-frequency connection of the microphone, the quality of the insulation can be lower than that of the widely used electrometric method of connection.

The stabilization of the supply voltage and the extreme negative feedbacks in the circuit ensure the necessary stability of the measurement.

TECHNICAL DATA OF THE INSTRUMENT

Measurement ranges 0–50, 0–80, 0–400 mg/m³.
Output signal 0–5 mA.
Basic error ±5% of the measurement range.
Reproducibility of the readings ±1% of the measurement range.
Duration of the transient process not more than 2 min.
Supply voltage 220 V, 50 Hz.
Power consumed not more than 250 W.
Weight not greater than 35 kg.
Ambient temperature 10–35°C.
Parameters of the analyzed gaseous mixtures:
    moisture content — not more than 1 g/m³,
    content of dust and other mechanical contaminants — not more than 0.001 g/m³,
    temperature between +10 and 35°C,
    pressure 680–785 mm of mercury.

To carry out the measurements in a mobile laboratory, a special mounting with shock absorbers was developed at the Main Geophysical Observatory (by Gal'dinov), which considerably reduced the vibration during car motion and made it possible to take measurements at the moment the car stopped. Squares and streets with dense traffic were selected as car stops for measurement purposes. The single (averaged over 20 min) concentrations of carbon monoxide fluctuated over wide limits (10–80 mg/m³) from point to point. These values are probably lower than the true peak values, since road safety considerations made it necessary to carry out the measurements rather to the side of the main flow of traffic.

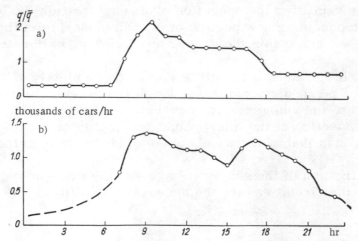

FIGURE 3. Daily cycle of the mean hourly concentrations of carbon monoxide (a) and intensity of traffic flow (b).

Helicopter measurements of carbon monoxide concentrations were carried out by a method already developed /5/. Methods for collecting air samples and for fastening the equipment in the helicopter cabin were worked out during the flights, and the concentrations of CO at heights of 100—300 m were measured. When carbon monoxide was recorded under stationary conditions, the gas analyzer was placed in a special enclosure. Air samples were taken through a special opening. The measurements were recorded on a 0—40 mg/m^3 scale. The servicing of the instrument consisted in adjusting the nul point and in changing the paper strip.

Figure 3a shows the diurnal cycle, averaged over 1.5 months (June—July 1970), of the mean hourly concentrations of carbon monoxide at one of the town squares, where traffic was the main source of pollution. (Here q is the mean hourly concentration, $\bar{q}$ the mean diurnal concentration.) At night there was almost no traffic, and the concentrations of carbon monoxide are near to zero. Traffic increases in the morning hours (see Figure 3b) and reaches a peak at about 9 o'clock. During that period the weakly developed turbulent exchange also contributes to increase in the concentrations. The maximum concentrations are observed at 9—10 hours. On the curve of traffic intensity a second peak appears in the afternoon, probably because of the end of the working day. However, due to meteorological conditions, the concentration of carbon monoxide does not increase in that period. The maximum value of the mean hourly concentration of CO is equal to 6 mg/m^3, on the average; concentrations of up to 15 mg/m^3 were observed on some days in the morning.

The results plotted in Figure 3 correspond to the working days of the week. During the weekend the concentrations are much lower, and usually fluctuate between 0 and 2 mg/m^3, attaining 3—4 mg/m^3 in particular cases.

We note, in conclusion, that tests on the gas analyzer GMK-3 under field conditions confirmed its highly satisfactory performance and yielded a number of additional data on the distribution of carbon monoxide in the air of cities.

Bibliography

1. Alekseeva, M.V.   Opredelenie atmosfernykh zagryaznenii (Determination of Atmospheric Pollutions).   Moskva, Medgiz. 1963.

2. Al'perin, V. Z. et al.   Avtomaticheskii gazoanalizator dlya
        nepreryvnogo opredeleniya sernistogo gaza v atmosfer-
        nom vozdukhe (Automatic Gas Analyzer for Continuous
        Determination of Sulfur Dioxide in Atmospheric Air). —
        Trudy GGO, No. 234, pp. 175—180. 1968.
3. Gal'dinov, G. V. and S. A. Kon'kov.   Nekotorye rezul'taty
        avtomaticheskoi registratsii sernistogo gaza i pyli
        (Some Results of the Automatic Recording of Sulfur
        Dioxide and Dust). — Trudy GGO, No. 238, pp. 96—106. 1969.
4. Gorelik, D. O. and B. B. Sakharov.   Optiko-akusticheskii
        effekt v fiziko-khimicheskikh izmereniyakh (The Optical-
        Acoustic Effect in Physicochemical Measurements).
        Moskva. 1969.
5. Goroshko, B. B., A. S. Zaitsev, and V. Ya. Nazarenko.
        Voprosy metodiki i rezul'taty issledovaniya zagryazneniya
        atmosfery s pomoshch'yu vertoleta (Problems of the
        Procedure, and Results of the Study of Air Pollution by
        Means of a Helicopter). — Trudy GGO, No. 234, pp. 85—94.
        1968.
6. Control Techniques for Carbon Monoxide Emissions from
        Stationary Sources. — National Air Pollution Contr. Admin.
        Publ., No. AP-65, Washington, March. 1970.

EXPLANATORY LIST OF ABBREVIATIONS
APPEARING IN THIS BOOK

Abbreviation	Full name (transliterated)	Translation
AANII	Arkticheskii i Antarkticheskii Nauchno-Issledovatel'skii Institut	Arctic and Antarctic Scientific Research Institute
AN SSSR	Akademiya Nauk SSSR	Academy of Sciences of the USSR
DAN SSSR	Doklady Akademii Nauk SSSR	Reports of the Academy of Sciences of the USSR
GGO	Glavnaya Geofizicheskaya Observatoriya im. A. I. Voeikova	Voeikov Main Geophysical Observatory
GMTs	Gidrometeorologicheskii Nauchno-Issledovatel'skii Tsentr	Hydrometeorological Research Center
GruzSSR	Gruzinskaya SSR	Georgian SSR
MGU	Moskovskii Gosudarstvennyi Universitet	Moscow State University
TsAO	Tsentral'naya Aerologicheskaya Observatoriya	Central Aerological Observatory
TsIP	Tsentral'nyi Institut Prognozov	Central Weather Forecasting Institute
VNIIOT	Vsesoyuznyi Nauchno-Issledovatel'skii Institut Okhrany Truda VTsSPS	All-Union Scientific Research Institute of Work Safety of the All-Union Central Trade-Union Council